Introduction to
CHROMATOGRAPHY

Second Edition

Roy J. Gritter

I B M Corporation
San Jose, California

James M. Bobbitt

Department of Chemistry
University of Connecticut, Storrs

Arthur E. Schwarting

Dean Emeritus
School of Pharmacy
University of Connecticut, Storrs

 Holden-Day, Inc.
Oakland, California

PREFACE

In our preface to the first edition (1968), we stated that chemistry is becoming even more dependent on chromatography as an ultimate method for the separation of mixtures. Over the past sixteen years this dependency has accelerated, and the field has changed in several ways.

This new edition has been completely reorganized and rewritten. Only a few paragraphs in Chapter 5, "Column Chromatography," remain the same. We have recognized that the field consists of two areas: gas chromatography (GC) and liquid chromatography (LC), and have organized the chapters accordingly. After Chapter 1, "Introduction and Theoretical Concepts," there is a chapter on GC and four on the various types of LC. Furthermore, we have reorganized our nomenclature along the same lines. Thus, the terms liquid-solid chromatography (LSC) and liquid-liquid chromatography (LLC) are used in place of adsorption and partition chromatography, etc.

The major advance in chromatography has been high performance liquid chromatography (HPLC). It has become an excellent method for obtaining qualitative and quantitative results and may well become the method of choice for preparative separations. We discuss HPLC in Chapter 6 as the ultimate method of liquid chromatography.

The short section on theory in the first edition has been expanded to show how a few simple measurements and calculations can be used to understand and expedite laboratory separations. Most of this is in Chapter 1, but bits are found in Chapters 2 and 6 on GC and HPLC, respectively.

As in the first edition, we devote most of our attention to the basic techniques involving solubility, adsorption, and volatility, and have given short shrift to such methods as ion exchange, ion pair, gel permeation, and many other important techniques. These are, however, defined in the glossary at the end of the book.

The numbers of books on chromatography continue to increase, and many are listed in the bibliography. Most of the suppliers of chromatography chemicals and equipment are included in the list of suppliers.

We have tried to provide an extensive index. Thus, there are multiple listings of the many topics and important words so that pertinent information can easily be found. In addition, the techniques involved for the various entries are most often given in the index. Thus, acid layers is entered as acid layers-TLC, etc.

We would like to repeat, from the Preface to the first edition, our basic goal for this book. It should contain enough information so that a beginning student (or chromatographer) can obtain, in fairly short time, an overall picture of chromatography. This must involve some simple answers and generalizations for questions that cannot be answered simply. We hope that our reader will respect our attempts and accept our generalizations as such and no more.

We are pleased to acknowledge the assistance of many of our chromatographer friends who have either read portions of the manuscript or who have carried out chromatographic separations for us. Readers include Kerry Nugent and Gary Adams of IBM Instruments and Professors Julian Johnson and James Stuart of the University of Connecticut. Sam Spencer of IBM Instruments produced all of the gas chromatograms. It is a credit to his skill that it took him the longest time to produce the example of the faulty injection shown in Figure 2.11. The HPLC chromatograms were produced by several scientists of IBM Instruments, and we appreciate the use of their data. Finally, we must acknowledge the use of the text processing capabilities of IBM Research-San Jose computer facility over the past few years. Jim Cox was especially helpful in the production of the proper format and camera ready copy, as well as finding the hidden flaws time after time. The backup capabilities of the computer were especially important in the saving of the GC chapter when it was inadvertently erased.

Roy J. Gritter
San Jose, California

James M. Bobbitt
Storrs, Connecticut

Arthur E. Schwarting
Cape Haze, Florida

CONTENTS

Chapter 1

INTRODUCTION AND THEORETICAL CONCEPTS

INTRODUCTION

The various methods of chromatography provide the most powerful separations techniques in the chemistry laboratory. The basic ideas are simple to grasp; the techniques vary from simplicity itself to fairly complex operations and instrumentation; and the methods are applicable to every type of substance. Although the word **chromatography*** implies color, there is no direct connection except that the first compounds separated by the technique were the green pigments of plants.

The chromatographic method, because of its broad utility, is widely used for analytical as well as preparative separations. Almost every chemical mixture from low to high molecular weight can be separated into its components by some chromatographic method. The type of separation, analytical or preparative, is not defined by the sample size, but more likely by the specific need. Normally, analytical chromatography would initially be used on all samples, and preparative chromatography would be carried out only when pure fractions of a mixture are needed.

Chromatographic separations are carried out by straightforward manipulations of certain of the general physical properties of molecules. The major properties involved are: (1) the tendency of a molecule to dissolve in a liquid **(solubility),** (2) the tendency for a molecule to attach itself to the surface of a finely divided solid **(adsorption),** and (3) the tendency for a molecule to evaporate or enter the vapor state **(volatility).** In a chromatographic system, the mixture to be separated is placed in a situation such that its components must exhibit *two* of these properties. This may involve two different properties such as adsorption and solubility, or it may involve one property in two environments such as solubility in two immiscible liquids.

Although chromatography is a *dynamic* interplay between properties, it can best be approached by first considering some *static* situations. For example, if one places a compound in a separatory funnel with two liquids

---

*The **bold type** will be used the first time an important term is mentioned. Other especially important concepts or ideas will be given in *italics*.

having a limited mutual solubility (such as ether and water), the compound or **solute** will tend to distribute itself or to **partition** between two liquids or **phases** depending upon its solubility properties (Figure 1.1a). This technique is, of course, the basis of simple extraction procedures. Such a partition represents a competition between solubility in two liquids. When one places a solute in a flask with a liquid and a finely divided solid (such as charcoal), the solute will distribute between the liquid wherein it exhibits a solubility property and the solid surface on which it will exhibit an adsorption property (Figure 1.1b). Finally, when one places a solute in a flask containing a small amount of a non-volatile liquid, the solute will distribute between the vapor state and solubility in the liquid, thus exhibiting solubility and volatility properties (Figure 1.1c). These systems can all be described as two phases in close contact and in equilibrium with one another and with a solute that is distributed between them.

It is highly unlikely that two compounds will exhibit *exactly* the same behavior with respect to the *two* phases. In a chromatographic system, it is possible to capitalize on such differences, even when they are very small, and to make them the basis for separations.

The basic idea in chromatography is to convert a static distribution situation as described above (and in Figure 1.1) into a flowing, dynamic equilibrium system. This is brought about by causing one phase (the **moving phase** or more correctly the **mobile phase**) to move mechanically with respect to the other phase (the **stationary phase**) while remaining in equilibrium with it. This is shown schematically in Figure 1.2a. (Although a

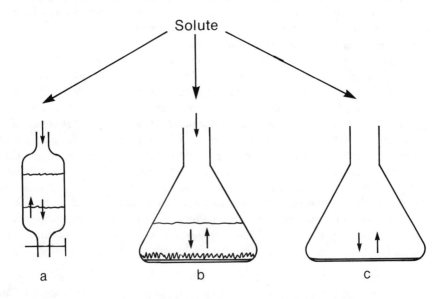

Figure 1.1 Static equilibrations of solute between two phases.

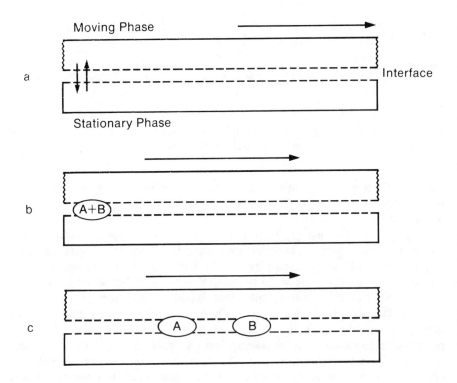

Figure 1.2 A schematic description of a chromatographic system without any solute (a), with two solutes at the beginning of a separation (b), and with two solutes after separation (c).

small space is shown between the phases in the figure for the sake of clarity, it should be understood that they are in direct contact with one another.) The mobile phase may be a liquid or a gas, and the stationary phase may be a liquid film on some type of support or a solid surface functioning as such. If the mixture to be separated (A plus B) is then introduced into the system (Figure 1.2b), the two compounds will distribute themselves between the two phases according to their respective properties. Since one of the phases is moving, the substances in the mixture must also move. The compound having the greater affinity for the mobile phase (or lesser affinity for the stationary phase) will move faster than the compound with the opposite properties.

After a period of time, the situation can be depicted as shown in Figure 1.2c, where compound B has moved at a faster rate than compound A. If A and B have very different properties, one may flow with the mobile phase and the other remain in the introductory position. In this case, the separation is easy. More usually, A and B have similar properties and both migrate, but at different rates. This difference in migration rates is the

		Moving Phase	
		Liquid	Gas
Stationary Phase	Solid	Liquid-Solid Chromatography or LSC	Gas-Solid Chromatography or GSC
	Liquid	Liquid-Liquid Chromatography or LLC	Gas-Liquid Chromatography or GLC

Figure 1.3 Types of chromatography showing how the name is derived from the nature of the mobile and stationary phases.

basis for all chromatographic separations, and the search for conditions that will produce the greatest difference between the rates represents the challenge of chromatography.

There are a confusingly large number of different types, variations, and techniques of chromatography. However, it is possible to make some general divisions and perhaps view the field in a reasonably systematic way. On the basis of the mobile phase, which may be a liquid or a gas, it is possible to divide chromatography into **liquid chromatography (LC)** and **gas chromatography (GC)**. On the basis of the stationary phase, which may be a liquid or a solid, it is possible to divide chromatography into **partition chromatography** or **adsorption chromatography.** In an obvious blend of these two concepts, one may then have, for example, liquid-solid chromatography, where the mobile phase is liquid and the stationary phase is solid. Other combinations are shown in Figure 1.3. We will use the terms in the figure.

The general division of the field into liquid chromatography and gas chromatography appears to be the more useful division and will be the one stressed in this book. A number of other types of chromatography have been devised that involve phenomena other than simple partition, adsorption and volatility. These types of chromatography will be briefly described in a Glossary of such methods in the back of this book. The Glossary also contains other names for the general types of chromatography.

1.1 DEFINITION OF TERMS

The introduction above has required the definition of a number of terms, but several more need to be defined for a general understanding of chromatography.

The solid phase, which serves as the stationary phase for liquid-solid or the less used gas-solid chromatography, is called the **adsorbent,** whereas the material that holds a stationary phase in liquid-liquid or gas-liquid chromatography is called the **support.** When the mobile phase is caused to flow over the stationary phase, to effect the chromatographic separation, the process is known as **development.** After the substances have been separated by development, the results are **detected** or **visualized.** When the substances being separated are actually washed out of the system, they have been **eluted,** or **elution** has taken place. The substances being separated are normally termed the **solutes,** or, collectively, the **sample.** The total result is then called a **chromatogram.**

1.2 LIQUID CHROMATOGRAPHY

A number of techniques have evolved in which the mobile phase is a liquid. These differ from one another in several ways. First, some involve a solid stationary phase (liquid-solid) and some involve a liquid stationary phase (liquid-liquid). These are often called **adsorption** and **partition chromatography,** respectively. Second, the techniques of LC differ in the shape or conformation of the stationary phase. The phase may be in a thin layer on some type of support as in **thin layer chromatography (TLC)** or as such in **paper chromatography (PC),** or it may be in a column that is held in place by a glass, metal, or plastic tube as in **column chromatography.** Finally, the techniques of LC differ in the rates at which the mobile phase is moved and in the methods by which the separations are detected. When the mobile phase is allowed to flow down through a column by a gravity flow, the method is called simply column chromatography. When the mobile phase is moved rapidly under pressure and the results are detected instrumentally, the process is called **high performance liquid chromatography, HPLC.**

A liquid chromatogram may be developed with a pure solvent, a mixture of pure solvents (both sometimes called **isocratic systems),** or, more often, a constantly changing mixture of solvents, usually two or three. When a changing mixture of solvents is being used, it is said that a **gradient development** or **gradient elution** is taking place.

Thin Layer and Paper Chromatography

These two techniques are similar in that the stationary phase is a thin layer and the mobile phase is allowed to flow through by capillary action. They differ in the nature and function of the stationary phase. In paper chromatography, the stationary phase is a liquid, usually water, suspended on the fibers of a piece of high-grade filter paper, thus giving rise to liquid-liquid chromatography. In TLC, the stationary phase is a thin layer (0.1-2 mm thick) of some solid material deposited on a flat supporting surface that is usually glass, but may be a polymer film or a metal foil. The layer is held in place by a binding agent of some type, usually plaster of Paris or starch.

In TLC the layer generally functions as a solid adsorbing surface (liquid-solid, see Figure 1.3), although it can also be used as a liquid support, giving rise to liquid-liquid chromatography. Of the two techniques, PC is the older, but probably TLC is much more commonly used at present. While it is possible to carry out a gradient development of a paper or thin layer chromatogram, the apparatus is complex, and simple development with a pure solvent or an unchanging mixture of solvents is more common.

The same series of operations is involved in PC and TLC, but they will be illustrated using a TLC system. The mixture to be separated is dissolved in any suitable solvent, preferably the developing solvent or one similar to it in polarity (see Chapter 3) and applied as a spot (1-5 mm in diameter) to the layer a short distance (ca. 2 cm) from one end. Such an application is generally made with a glass capillary (Figure 1.4), but can be made with a syringe or an automated device. The application solvent is allowed to evaporate or is removed in a stream of dry air or nitrogen. The layer is then placed in a developing chamber containing a layer of solvent about 1 cm deep which will act as the mobile phase. This is done in such a manner that the solvent is in contact with the layer on the end nearest the sample spot, but, of course, below it (Figure 1.5). The chamber is then closed securely, and the solvent is allowed to ascend the layer by capillary action to a point 10-15 cm above the sample spot (Figure 1.6).

Paper chromatography can be carried out in much the same manner except that the paper must be suspended from some type of hook in the chamber since it does not have a firm backing. When the mobile phase and the stationary phase have been properly selected, the original sample spot will have been resolved into a series of spots, each hopefully representing a single component of the mixture (Figure 1.7). The chromatography is usually carried out in a chamber that has been saturated as completely as possible with the mobile phase. This is the purpose of the piece of filter paper that partially lines the chamber in Figures 1.4 through 1.7. When the spots are not colored, they must be visualized by spraying them with a suitable color-developing reagent or by placing them under an ultraviolet light.

An almost infinite number of variations of this simple procedure have been published since PC and TLC were first described. Some of these are important; some are trivial; and some are complex beyond the scope of this introductory book. The important ones will be described in Chapter 4.

Several terms are used in TLC and PC in addition to those given earlier. The point at which the mixture is deposited at the beginning of the chromatogram is called the **origin,** and the technique of placing it there is known as **spotting.** The **solvent front** is the top of the mobile phase or **solvent** as it moves through the layer and, after the development is completed, represents the maximum height achieved by the solvent. The behavior of a specific compound in a specific chromatographic system is described

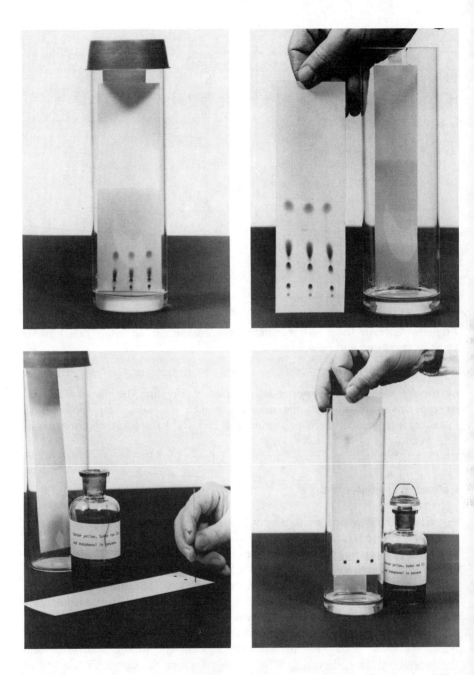

Figure 1.4-1.7 A mixture of three dyes in benzene is spotted on a silica gel G layer in three concentrations (1.4), placed in a saturated chamber containing benzene as a developer (1.5), developed (1.6), and removed and dried (1.7).

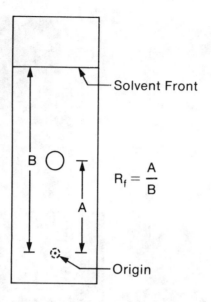

Figure 1.8 An idealized thin layer chromatogram showing how an **Rf** value is measured and calculated.

by the *Rf* **value.** This number is obtained by dividing the distance moved by the solvent front into the distance moved by the solute spot. Both distances are measured from the origin, and *Rf* values vary from 0 to 1. This is shown in Figure 1.8.

Column Chromatography

Classic column chromatography is the oldest of all of the many chromatographic methods and, as it has been traditionally practiced, is a form of liquid chromatography. The stationary phase, either an adsorbing material (LSC) or a supported liquid film (LLC), is placed in a glass cylindrical tube closed at the bottom by a valve or stopcock, and the mobile phase is allowed to flow down through it by gravity flow. The **chromatographic column,** Figure 1.9, is usually prepared by pouring a slurry of the stationary phase in a suitable solvent into the column and allowing it to settle out. Next, the level of the solvent is lowered to the top of the adsorbent, and the sample mixture, dissolved in a suitable solvent, is placed on the top of the column and allowed to flow into the top layer of adsorbent or support (Figure 1.10). The mobile phase solvent is then placed on top of the column and allowed to flow through and develop the chromatogram (Figure 1.11). Under well-chosen conditions, the solute components of the mixture

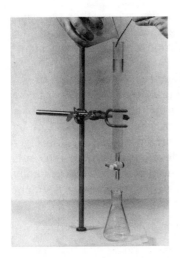
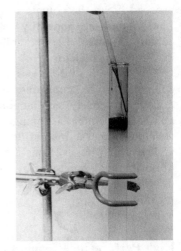

Figures 1.9–1.12 A slurry of silica gel in benzene is poured into a glass column (1.9) and the solvent is allowed to flow out until the level is just over the adsorbent. A small circle of filter paper is placed on the top of the column and the dye mixture (see Figure 1.4) is added with a dropper (1.10). Additional benzene, as a developer, is allowed to flow through the column so that the colored bands are resolved (1.11) and finally until one band emerges from the bottom (1.12). The time involved in this particular experiment was just about 30 min.

proceed down the column in **bands** at different rates and are thus resolved. The solutes are generally isolated by allowing them to flow out of the column (Figure 1.12) and collecting them as fractions, often with a mechanical fraction collector.

In the case of colorless compounds, the effluent from the bottom of the column must be monitored to find out where the solutes are. This can be done continuously with a suitable detector or by dividing the effluent into a number of sequential samples and analyzing them, generally by TLC, or by weighing the fractions after the solvent is evaporated. Whereas thin layer and paper chromatograms are more frequently developed with pure solvents or unchanging mixtures of solvents, columm chromatograms are generally developed with constantly changing mixtures of solvents by a gradient technique.

High Performance Liquid Chromatography

Theoretically, the best chromatographic separations will be produced when the stationary phase has the largest possible surface area, thus ensuring a good equilibrium between the phases. A second requirement for good separation is to have the mobile phase flowing rapidly to ensure a minimum diffusion situation. A large stationary phase surface area means, in most chromatographic situations, a finely divided adsorbent or support. In order to force a rapidly moving mobile phase through a finely divided stationary phase, high pressures must be used. These requirements have given rise to the newest and most powerful of the techniques of liquid chromatography. At first it was called **high pressure liquid chromatography,** abbreviated as **HPLC.** This name was modified to **high performance liquid chromatography,** still **HPLC,** and is sometimes, and incorrectly, referred to as **liquid chromatography** or **LC.** We will consistently use HPLC since we have defined LC in the more general sense as any chromatography that involves a liquid mobile phase.

Still another unique aspect of HPLC is the use of very sensitive **detectors** of one sort or another to analyze the effluent from the column when colorless or very low concentrations of solutes are being separated. These detectors may involve a continuous monitoring of the ultraviolet absorption, the refractive index, or some other physical constant of the effluent that will change sufficiently as the solutes emerge from the column. In short, when some of the advances that were developed for gas chromatography were applied to classical chromatography, the technique of HPLC was born.

HPLC is carried out either as an liquid-solid method or as a liquid-liquid method. The LLC technique uses either a stationary phase *chemically bonded* to the support or, less often, adsorbed to the support. Both chromatographic methods use a packing or support that is very finely divided (3-20 μm). For analytical purposes, such a packing is put into a

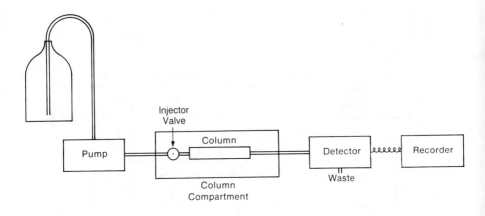

Figure 1.13 Schematic diagram of an apparatus used for HPLC.

stainless steel tube of small diameter (2-6 mm) and moderate length (5-30 cm). The mobile phase is pushed through the column at pressures ranging from 20-10,000 psi. For preparative separations, larger column diameters are used.

Figure 1.13 is a diagram of an apparatus used for HPLC. About half of the parts have some similarity to those used in gas chromatography. The various parts are: an inert, gas-vented storage container for the liquid phase, a high pressure solvent delivery unit, an injection system or valve, an oven (rarely thermostated), the column itself, a detector and associated electronics, and a recording device. All of the connecting tubing and valve systems are designed as small and short as possible to minimize **extra column volume.** Gradient elution or solvent programming techniques are often used in HPLC. That, and the need for a very consistent and surgeless solvent flow, is the reason for the complex solvent delivery unit.

The chromatographic system is operated in the following manner (see Figure 1.14). The mobile phase is forced through the column under the desired pressure and at the desired rate. After the system has reached equilibrium, the sample dissolved in a suitable solvent is injected into the system, generally through a valve. The solutes are carried into the column, separated, and pass out in the effluent through the detector. The detector results are plotted on chart paper, and the resulting graphs can be used to obtain analytical data or to find out where the various mixture components are in the effluent. Note that the sample components are in the form of a **band** or **plug** when they are first placed in the system and that they broaden as they flow through the system.

The separation of a particular mixture of solutes by HPLC under a given set of conditions is quite repeatable, and it is possible to assign numbers to the solutes that describe their behavior. These numbers are

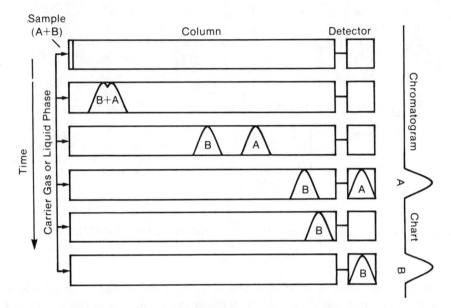

Figure 1.14 A more detailed schematic diagram of GC and HPLC. Solutes A and B can be seen to separate as they pass through the column and the results are recorded on a chart as the components emerge from the column.

called **retention volumes** and are a measure of the volume of solvent needed to move a specific compound through a given system. The number is, in a way, analogous to the Rf value as it is used in TLC and PC. The retention volume concept will be discussed in greater detail and more carefully defined later.

1.3 GAS CHROMATOGRAPHY

In gas chromatography, or GC, the mobile phase is an inert gas such as helium, nitrogen, argon, or even hydrogen which is passed under pressure through tubing containing the stationary phase. Although gas-solid chromatography, wherein the stationary phase is a solid surface, is well known, it is much less common than gas-liquid chromatography, where the stationary phase is a liquid film. For chromatographic separations, the liquid stationary phase is present as a thin coating adsorbed or chemically bonded to a solid support which is, in turn, packed in a small diameter (2-8 mm) metal, glass, or plastic tube of moderate length (1-10 m). This is called a **packed column**. In an alternate system called a **capillary** or **open-tubular column**,

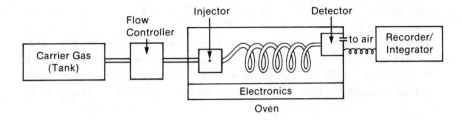

Figure 1.15 A schematic GC apparatus. The sample is injected into the carrier gas stream and allowed to flow into the column where it can be partitioned between the gas phase and the liquid-coated support. The sample is separated and emerges from the column into the detector.

the stationary phase is present as a thin film (0.1-2 μm) on the inside wall of a glass or metal capillary of very small diameter (0.2-1 mm) and very long length (10-100 m).

The column is placed in a thermally controlled oven that can be cooled or heated. Since one of the two key properties exhibited by solutes in GC is solubility, and solubility (or volatility) is closely related to temperature, a precise control of temperature is mandatory. We stated earlier that a major advantage of LC was that the constitution of the mobile phase could be continuously changed (gradient elution) in order to facilitate separations. In an analogous fashion, the temperature can be continuously changed during GC to increase, systematically, the volatility of the solutes and thus enhance separation. In such an operation, the temperature is said to be programmed and the result is **temperature programming** as compared to **gradient elution** in HPLC.

Figure 1.15 shows, schematically, a GC apparatus consisting of a few basic parts, many of which are analogous to those shown in Figure 1.14 for HPLC. The components are a high pressure purified gas supply with a pressure regulator, a sample inlet system or **injector,** a thermostated oven, a column with a suitable packing, a detector and associated electronics, and a recording device for the detector. As in HPLC, it is desirable to keep the non-volume as low as possible. The system is operated in a manner directly analogous to the operation of HPLC as described above and illustrated schematically in Figure 1.14.

The behavior of a specific compound under a given set of circumstances (column, flow rate, temperature) is quite characteristic. Thus, the compound will appear at the detector at a given time after it is **injected.** This is usually called the **retention time** and is the same as the retention volume defined above in HPLC. In actual fact, many use retention times to define both GC and HPLC. As long as the HPLC is done at a constant flow rate, the values represent the same thing.

1.4 APPLICATIONS OF CHROMATOGRAPHY

One may carry out chromatography to answer essentially three questions: What is present? How much is present? How do we get some of the pure components? We will consider the answers to these questions to be **qualitative, quantitative,** and **preparative chromatography,** respectively.

Qualitative Applications
(What Is Present?)

First of all, qualitative applications of chromatography reveal the presence or absence of a specific compound in a sample. In TLC and PC, this is generally done by comparing the pure compound with the mixture. In GC and HPLC, one compares retention times or volumes of the pure compound with those of the components of a mixture. In order to be detected in a given mixture, the suspected compound must be present in sufficient quantity to be measured (the **detectability level).** The retention time or retention volume will give an indication of the identity of the compound. In any case, conclusions from these simple operations should be confirmed by other methods, either chromatographic or more preferable spectrometric.

Secondly, qualitative chromatography gives information on the complexity of a mixture. The mixture is chromatographed under various conditions and even by several techniques or combined methods. The number of spots (TLC and PC) or peaks (HPLC and GC) will indicate the minimum number of mixture components. The number must be considered to be a minimum number since it is not possible to prove that any given spot or peak is a pure material rather than a mixture. Conversely, the purity of a given compound can be studied. In this case, the compound is chromatographed under various conditions and concentrations (up to the capacity of the system). The presence of one TLC or PC spot or one GC or HPLC peak is a good criterion of purity at a stated detectability level and is frequently cited as such.

Finally, qualitative chromatography can often be used to establish a component fingerprint pattern for complex mixtures in which the components may be known or only partially known. These may be established on such mixtures as tissue extracts, urine, blood, a crude chemical, or a drug. Samples to be investigated can be chromatographed and the results can be compared to the normal pattern.

Although all of the chromatographic methods, except perhaps column chromatography, can be used qualitatively, the various methods have certain advantages. TLC and PC are the simplest and require less expensive equipment and they also permit multiple samples in a single run. TLC is much faster than PC. Both are unsatisfactory for volatile compounds. GC and HPLC require complex equipment, but have an extremely high resolving power. In general, GC is used for the investigation of liquid and volatile materials (b.p. up to 400°C), and HPLC is usually used for soluble solids and less volatile liquids. In some cases, volatile compounds are converted to non-volatile compounds for separation by HPLC or TLC and, conversely, non-volatile compounds are converted into volatile compounds for separation by GC. The technique of choice often depends on the available equipment.

The two major advantages of chromatography as a qualitative method are that very small samples are required for analysis and that the analyses normally require short times. The lower limits on sample size are governed only by the sensitivity of the detection system used and are, at present, about 1×10^{-8} to 1×10^{-12} g for GC, 1×10^{-6} to 1×10^{-10} g for HPLC, and 1×10^{-4} to 1×10^{-8} g for TLC. Most analyses can be carried out in less than half an hour.

Different types of scientists will seek qualitative information of various sorts. The synthetic chemist, organic or inorganic, will examine reaction mixtures to determine which conditions give the cleanest products and which reactions do not occur at all (giving only starting materials). Furthermore, he might remove samples from a reaction at timed intervals for chromatography. This type of experiment can reveal information about possible intermediates and optimum reaction times. A chromatogram that might result from such an experiment is shown schematically in Figure 1.16.

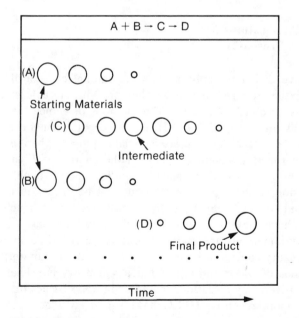

Figure 1.16 A kinetic study by TLC of the reaction between A and B to yield the intermediate C that is, in time, converted to the final product D.

After the reaction is completed, one may separate the reaction mixture into various fractions by a variety of techniques (including chromatography) and inspect each of the fractions to learn where the product is.

The natural product chemist, pharmaceutical chemist, medicinal chemist, or biologist can monitor all of the steps in a complex separation scheme down to the final purification. The qualitative (and quantitative) examination of tissues, blood, and urine for the presence of various poisons or drugs is an important aspect of toxicology and forensic and clinical chemistry. Fingerprint chromatograms of mixtures of components, described previously, are used extensively in industry for quality control and have been used in medicine to study the fate of drugs after they have been administered. Patterns have been used to look for the presence of odd metabolites that may be present in certain diseases or conditions.

Qualitative chromatography is being used extensively by analytical chemists for studies of air and water pollution; agricultural chemicals (herbicides, insecticides, fungicides, and fertilizers) and their residues; foods and food additives; industrial chemicals and their impurities; pharmaceuticals; essential oils; hydrocarbons and petroleum (the chromatogram of an oil sample will indicate its source); polymers and their decomposition products; carcinogens; and a multitude of other materials. Chromatography is considered to be a branch of analytical chemistry, and analytical chemists have generally been responsible for most of the advances and refinements in the various techniques.

Quantitative Applications
(How Much Is Present?)

Quantitative chromatography reveals the amount of each of the components of a mixture relative to one another or, with suitable standards and calibrations, as absolute quantities. The methods of choice for quantitative chromatography are surely GC and HPLC, since the results are plotted out on chart paper by a recorder or plotter or data system. The areas under the peaks in the chromatogram that appear on the chart (see Figure 1.14) can be integrated by a variety of methods depending upon available equipment, and the precision required. These methods will be described more fully in Chapters 2 and 6. TLC can be used quantitatively with much less precision, but does have the advantages of simplicity and low cost.

Quantitative methods are used for routine assays of samples, generally as a part of quality control in industry and especially in monitoring environmental problems of water and air. Clinical chemistry is primarily a matter of the quantitative assay of various materials in body tissues or fluids.

GC and, more recently, HPLC have produced a major revolution in organic chemistry since the mid-1950's. These methods have made possible the facile and accurate measurement of product distributions in reaction mixtures.   Such data have, of course, aided synthetic chemistry, but its major use has been in the study of reaction mechanisms and physical organic chemistry.   The special advantage of chromatography in this context is that reaction mixtures can often be analyzed directly without any prefractionations which might change the accuracy of the results.

**Preparative Applications
(How Do We Get Some?)**

Preparative chromatography is used to obtain reasonable amounts (mg to g) of mixture components in a pure state so that they may be more completely characterized or used in further reactions.   Column chromatography was originally devised for this purpose and has been used extensively for many years.   Preparative TLC carried out on layers as thick as 1 cm **(thick layer chromatography)** has its usual advantages of simplicity and low cost.   The liquid-liquid columns that are usually used in GC and HPLC have fairly low capacities and cannot be used for the separation of large quantities in a single chromatogram.   However, ingeneous multiple injection systems have been devised to partially alleviate the problem, and large column assemblies are also available.

The major advances in preparative chromatography have, however, come about as the various technical developments were applied to column chromatography to give HPLC.   These technical developments, such as smaller and more homogeneous particle size, bonded phases, and continuous monitoring of column effluent are continuing and, when complete, will surely cause HPLC to be universally accepted as the ultimate preparative method (until a better one comes along).

Preparative separations are daily occurrences in all fields of chemistry, biology, pharmacy, and medicine.   Consider almost any isolation or identification problem undertaken today and there is a high probability that preparative chromatography was used in one way or another.

THEORETICAL CONCEPTS

A number of approaches are available to chromatographic theory.   We will use the theoretical plate concept.   This was the basic approach used by Martin and Synge[1] in the development of liquid-partition chromatography, for which they received the Nobel Prize.   Our treatment will be fairly superficial;   a more precise discussion is given in several of the general chromatographic texts listed in the Bibliography in the back of the book, especially in the book by Snyder and Kirkland.[2]   A somewhat different approach can be found in a text by Laitinen.[3]

The Static Distribution

In the introduction to this chapter, we described the various distributions or partitions that take place when a solute is introduced into a two phase system (Figure 1.1). We then blithely stated that chromatography resulted when such a static situation was converted into a flowing system by one phase being caused to move mechanically with respect to the other. Before we consider the details of this process, there is one more aspect of the static distribution that we should consider. This aspect involves the response of such a static equilibrium to *increasing amounts of solute.* In order to measure such a response, we can carry out the following experiment. Suppose we introduce a small amount of solute into one of the systems shown in Figure 1.1, allow the solute to equilibrate, and measure its concentration in the two phases. Then, suppose we introduce *more* solute into the same system, allow equilibration to occur, and measure the concentrations in the two phases again. When this process is repeated and the concentrations in the two phases are plotted against one another, as shown in Figure 1.17, we will get a series of points that, when connected, will provide a **sorption isotherm.** It is understood, of course, that the amount of solute added will not exceed the capacities of the phases to dissolve or adsorb it.

One might think that the response of a system to increasing solute would be consistent and that the isotherm would be a **linear sorption isotherm,** as shown in Figure 1.17a. This is true in most liquid-liquid systems, especially when small amounts of solute are involved. Liquid-solid systems are complicated by the fact that the sorption isotherms are usually *not linear.* They may be **convex sorption isotherms** (Figure 1.17b) or **concave sorption isotherms** (Figure 1.17c), and it is impossible to predict

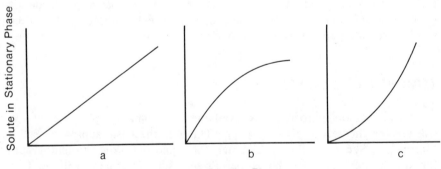

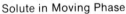

Figure 1.17 Various sorption isotherms: **(a)** a linear sorption isotherm reflecting an ideal distribution response to increasing amounts of solute, **(b)** non-linear and convex isotherm, **(c)** a non-linear and concave sorption isotherm.

which will be the case with any specific solute in any specific system. In fact, situations are common in which one solute will have a concave isotherm and another solute will have a convex isotherm in *the same system*. Because of these complications, our theoretical discussions will be confined to liquid-liquid systems having linear sorption isotherms. The effect of these deviations from linearity on peak shape will be considered in the later chapters of the text, mainly in connection with specific techniques.

An Intermittent Distribution Model

For our theoretical discussion, the conversion of a static system into a flowing system can be visualized as the placing of a *number of single distributions in series with one another,* as shown on the left side of Figure 1.18. In the series, the mobile phase is allowed to equilibrate with the stationary phase in each distribution. The valves are then opened and the mobile phase is allowed to flow onto the next distribution. In Figure 1.19, the two phases are liquids and the same amount of each phase is present in each distribution. In this case the upper phase is the mobile phase, but this is only a matter of mechanical design. When the various equilibration stages are then stacked upon one another, as shown on the right side of Figure 1.18, with no valves between them and with constant flow of mobile phase, we will have a reasonable approximation of a chromatographic system. Our theoretical model will thus consist of a series of discrete distributions. An apparatus has been developed that is analogous to the described model, complete with discrete distributions and liquid transfers. This is the Craig countercurrent apparatus, which is commercially available and used in some laboratories.

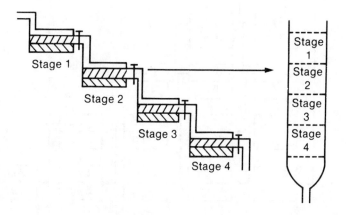

Figure 1.18 A schematic approach showing the conversion of an intermittant partition system into a continuous column chromatogram.

The behavior of a solute in such a liquid-liquid system will depend upon its relative solubilities in the two liquids involved. This is described in physical chemistry as the **partition coefficient** which we will designate as K. The partition coefficient is defined as the concentration of solute in one phase divided by the concentration of solute in the other phase under equilibrium conditions. K is easy to measure experimentally in a separatory funnel and should be constant at any given temperature (thus giving rise to the linear sorption isotherm seen in Figure 1.17a). In chromatography, K is defined as the concentration of solute in the stationary phase divided by the concentration of solute *(S)* in the mobile phase, as shown in equation 1.1.

$$K = \frac{S_{stationary\ phase}}{S_{mobile\ phase}} \tag{1.1}$$

Suppose that we first consider the behavior, in our model system, of a solute having a K value of 1, that is, a situation in which a solute will distribute itself equally between two liquid phases having equal volumes. Suppose that we then introduce 64 µg of solute into the first stage or stage 1 of a system shown in Figure 1.19 (64 makes the calculation easy, and the µg is a typical amount used in GC and HPLC). After equilibration, 32 µg will be in each phase (step 1 in Figure 1.19). The valve is then opened and the mobile phase is allowed to flow to stage 2 (see arrows in figure). Simultaneously, a fresh portion of the mobile phase is added to stage 1.

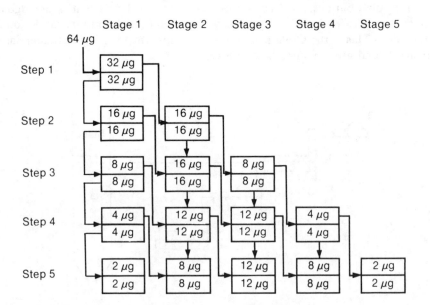

Figure 1.19 Solute distributions that will result after various numbers of equilibrations when the partition coefficient (K)=1.0.

After equilibration, the amounts in the phases will be as shown in step 2.   If the process is continued in the same manner through five steps, the solute will be distributed as shown, with the maximum amount in the middle stage. Note that, in the process, each distribution receives solute from two places, its own stationary phase from the previous step and from the mobile phase of the previous stage in the previous step (see arrows).   The total solute in each distribution is then partitioned according to the K value.   In Figure 1.20, the total amount of solute in each stage (from both phases) is plotted against the number of each stage.   The data from Figure 1.19 appear as the solid line.   If the system is extended to nine stages and seventeen stages, two additional curves (dotted) for nine and (dashed) for seventeen will result.

Four important conclusions can be drawn from Figures 1.19 and 1.20. First, a solute *must* ultimately be distributed in a number of stages of a system.   Second, the distribution of solute in the various stages will be a classical Gaussian distribution (note the shape of the peaks in Figure 1.14). In fact, the mathematics associated with a Gaussian curve is used to predict chromatographic properties.   Third, a solute with a K value of 1 will be concentrated at the center of the system.   The final, and probably most important aspect of the data, is that *when the number of stages is increased, the solute is concentrated in a smaller fraction of the total system.*

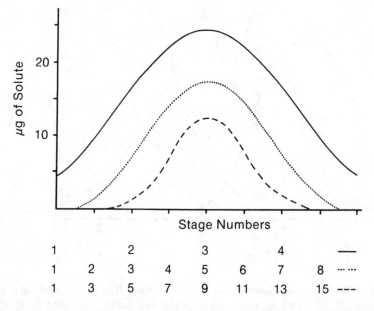

Figure 1.20 The distribution of a single solute after five (solid line), nine (dotted line), and seventeen (dashed line) equilibrations when the partition coefficient (K)=1.0.

Now, suppose that we consider the behavior of a mixture of two solutes in a similar system of discrete equilibria. Let us assign K values of 0.33 and 3.0 to two solutes A and B, respectively, and treat each solute with no regard to the other. If we begin with 64 μg of each solute, the results, after the same five (solid), nine (dotted) and seventeen (dashed) equilibrations are shown in Figure 1.21. Note that five equilibrations produce essentially pure solute only in the first and last distribution and that the total amount of each pure solute is only 21 μg (out of 64 μg). After seventeen equilibrations, the separation is greatly improved, and a total of about 61 μg of

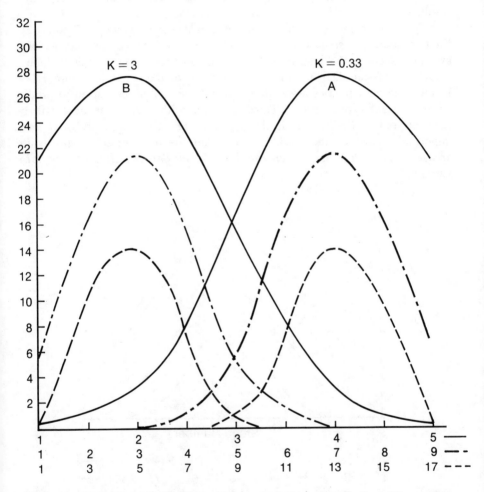

Figure 1.21 The distribution of two solutes, A and B, with partition coefficients of 0.33 and 3, respectively, after five (solid line), nine (long-short line), and seventeen (dashed line) equilibrations. A mixture consisting of 64 μg of each solute, 128 μg in all, has been separated.

essentially pure A can be found in distributions 1-8. A similar amount of essentially pure B is found in distributions 10-17. In this precise and predictable manner, chromatographic separations are produced, at least theoretically.

The experimental results shown in Figures 1.20 and 1.21 show clearly how a number of discrete, equilibrium stages will produce a separation of two solutes having different K values. Although true chromatography is continuous, it can, as stated previously, be considered as a series of theoretical equilibrations. These equilibrations are called **theoretical plates,** a term borrowed from distillation theory. A given column or thin layer of a specific length, then, has the ability to produce a separation *equivalent* to a certain number of equilibrations or theoretical plates. When the length of the column or layer is divided by the number of theoretical plates, one obtains a more useful number called the **height equivalent** to a **theoretical plate** or the **HETP.** The HETP is a measure of the efficiency of a given system, with smaller HETP's meaning a more efficient system. A good GC column may have a HETP of 0.5 mm/theoretical plate, which would mean that a 1 m column would produce separations equivalent to *2000 distributions*. A corresponding HPLC column with a typical HETP of 0.05 mm would have 20,000 theoretical plates.

Theoretical Models of Chromatograms

True chromatography differs from our model in several ways. First, chromatography is continuous with no discrete equilibration steps as such. This means that solute will *diffuse* through the system, both forward and backward, from a given theoretical position. This diffusion will result in zone or peak broadening. The problem of diffusion will be considered qualitatively in the section following this one.

Another difference between our model and a true chromatogram involves the amount of stationary phase present in relation to the mobile phase. In our model we assumed that the amount of mobile phase in each equilibration was equal to the amount of stationary phase. This situation is rarely, if ever, true in chromatography, for the amount of mobile phase in any cross section of a column or layer is always larger than the amount of the stationary phase. This fact requires a correction of the K values of the solute if they are to be used in our discussions. This correction factor is the ratio of the volume of mobile phase divided by the volume (V) of stationary phase in the same column or space and is designated as β (equation 1.2).

$$\beta = \frac{V_{mobile\ phase}}{V_{stationary\ phase}} \tag{1.2}$$

The β value is called the **phase ratio** and is a measure of the openness of a system. For example, a high phase ratio means that a relatively large

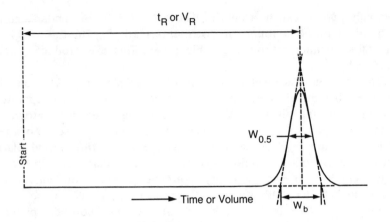

Figure 1.22 A theoretical peak from a GC or HPLC chart showing the various parts as they are named and used.

amount of mobile phase is present. Consequently, a large amount of space is available for the mobile phase to pass through, and the column is relatively open. The β value for a typical packed GC or HPLC column may vary from 5 to 35. For a capillary GC column, β varies from 50 to 1500. The β value for any system is usually obtained by measuring the amounts of the two phases used in the preparation of a column.

Still another difference between our model and real chromatography is the retention of the solutes within the model system. In some types of chromatography, the solute is removed or eluted from the system. In the model described in Figures 1.20 and 1.22, just enough fresh mobile phase was introduced to fill the system, regardless of the number of equilibrations, and the solutes, of course, remained within the system. This is true for thin layer and paper chromatography. This fact is not true of gas chromatography, column chromatography, and high performance liquid chromatography. In these methods, fresh mobile phase is continually added to the system, and the solutes are eluted out of the end of the column.

Let us first consider the cases of PC and TLC. In our previous discussion, we defined the Rf value (Figure 1.8) and stated that it was used to describe the properties of a solute in either of these two techniques. In the figure, Rf was defined in units of length, as it is usually measured. The Rf, however, is more correctly defined as the rate of motion of the solute divided by the rate of motion of the mobile phase. In PC and TLC, the solute and mobile phase migrate for the same time; the distances are proportional to the rates and the two definitions are equivalent. The Rf is related to the K value for any solute by equation 1.3.

$$R_f = \frac{1}{1 + K/\beta}$$

(1.3)

In the models described in Figures 1.20 and 1.21, the amounts of mobile and stationary phases were equal, therefore β was 1. For the solute in Figure 1.20 having a K value of 1,

$$Rf = \frac{1}{1+1} = 0.5$$

As is clear in the figure, the maximum amount of solute is present in the stages that are exactly halfway through the system, corresponding to the Rf of 0.5. For real cases in PC and TLC, β is not 1 and the relationship between Rf and K is harder to determine.

Now, suppose we consider the situation that arises when the solute zones are eluted from the column as is the case in GC, HPLC, and column chromatography. These can be called **elution** or **flowing chromatograms.** For this discussion, we will need to define more terms. The first of these is the **holdup volume** or **void volume.** This is the amount of mobile phase needed to surround the packing and fill the pores in a column. Put another way, it is the amount of mobile phase needed to move a cross section through the length of a column. The holdup volume is designated as V_M and can be measured in several ways. In a simple slurry packed column (Figure 1.9), V_M is equal to the initial liquid used to make the slurry, minus the liquid that flows out of the column as the slurry settles out and the level is allowed to fall to the top of the packed column. Probably the best way to measure V_M in a liquid system is to place a dye in the system that has no solubility in the stationary phase and that then will move at the same rate as the mobile phase. The volume of mobile phase required to move the dye through the system is then equal to V_M.

We noted in our discussion of Figure 1.20 that, as a consequence of the chromatographic process, any solute is present in several distribution stages, and that this distribution follows a Gaussian curve. This Gaussian distribution is, of course, still present as solute zones are eluted from the column. In GC and HPLC, it is customary to pass the column effluent through some type of detector that measures the amount of solute as it emerges. These data are plotted by a recorder against time (volume), and a strip chart describing the chromatogram is the result (see Figure 1.14). The solute zones, as they emerge, are usually called **peaks,** and one speaks of sharp or broad peaks.

Two other terms that we must consider were also defined previously for GC and HPLC. These are the **retention time** and the **retention volume.** The retention time is designated as t_R and it is used in both GC and HPLC. The retention volume is designated as V_R and it is used in discussions of HPLC and column chromatography. Both the retention time and the retention volume are measured to the *center* of the zones or peaks as they emerge from a chromatogram. We will use the volume terms in our discussion.

The quantity V_R is characteristic of any solute in a given system and can be calculated from the K value of the solute if the characteristics of the column V_M are known, equation 1.4. V_R can also be defined as V_{stat} in terms of the volume of the stationary phase.

$$V_R = V_M(1 + K/\beta) = V_M + KV_{stat.} \tag{1.4}$$

One might think, at this point, that solutes having different retention volumes would be separable and that we could now predict how the separation might take place. Unfortunately, the situation is not quite that simple.

Retention volumes are measured to the center of solute peaks. If the retention volumes are not sufficiently different or if the peaks are too broad, extensive overlapping will occur and poor separations will result. This fact is clearly shown in the five-distribution curve in Figure 1.21. The answer to the problem is also shown in Figure 1.21. The peaks become sharper and more separated when the column or system contains more distributions or theoretical plates. That is to say, a more efficient (more theoretical plates, low HETP) system will produce better separations (more **resolving power**). In our treatment, then, we need to consider the number of theoretical plates. The number of theoretical plates in a system is a function of the peak broadness and the retention volume of a given solute and is designated as N. Primarily from the mathematics of simple Gaussian distributions, N can be defined as shown in equation 1.5, in terms of the **peak width** at the baseline, W_b, or the peak width halfway up the peak, $W_{0.5}$. These values and the ways that they may be designated are shown in Figure 1.22 for a theoretical peak as it might appear on a strip chart from an HPLC or GC experiment.

$$N = 16\left(\frac{V_R}{V_b}\right)^2 = 5.54\left(\frac{V_R}{V_{0.5}}\right)^2 \tag{1.5}$$

The dotted line represents the way the peak might actually appear after some diffusion. The solid line represents the theoretical triangle associated with the peak. Note that the $W_{0.5}$ is a better approximation of what actually happens, since it is measured at the inflection point between a predicted curve and a probable curve. In the calculation of N from a peak on a chromatogram, $W_{0.5}$ would therefore be more valid. On the other hand, W_b is more useful if one wishes to predict the appearance of a given chromatogram from K values and a known N. The speed of the strip chart recorder is normally correlated with the rate at which the mobile phase passes through the system, or with the time needed for the various peaks to appear. Thus, the horizontal measurements $(V, t_R, V_R, W_b,$ and $W_{0.5})$ in Figure 1.22 can be expressed in terms of volume or time. We will use volume terms consistently for now.

We are now in a position to carry out some meaningful calculations on a system and to relate fundamental properties of solutes to their chromatographic properties. Suppose we make our measurements on an HPLC column packed with some support covered with a liquid film, either deposited or chemically bonded to it. By definition, we will assign a phase ratio of 5 (equation 1.2) and a holdup volume (V_M) of 0.40 mL. These quantities can be measured during column preparation and with a marker dye, as previously discussed. Such a column might be a stainless steel cylinder about 10 cm long and 0.4 cm in diameter.

First, we will need to find out how many theoretical plates are in our column. This can be done by chromatographing any solute that behaves in a reasonably theoretical fashion in the system, for example, one having a linear sorption isotherm. The nature of this solute is immaterial, since the value of N should be, and usually is, independent of the solute. From the strip chart, we can measure V_R and $W_{0.5}$. Note that V_R is measured from the point of injection. Suppose that V_R of our solute is found to be 4.0 mL and the $W_{0.5}$ is 0.40 mL. These data are recorded in the top part of Table 1.1 and are in **boldface** type, since they are obtained from actual *measurements*. The N can be calculated from equation 1.5 as follows

$$N = 5.54 \frac{(4.0)^2}{(0.40)^2} = 5.54 \frac{16}{0.16} = 554$$

and is recorded in Table 1.1 in *italics,* since it is calculated.

Using the N value calculated above, along with a measured phase ratio β, measured holdup volume V_M, and the measured partition coefficients of two solutes C and D (100 and 50), we can now begin to calculate and predict the exact theoretical shape of the strip chart record of the chromatogram. From equation 1.4, the retention volumes can be calculated.

$$(V_R)C = V_M(1 + K/\beta) = 0.4(1 + 100/5) = 8.4$$

$$(V_R)D = 0.4(1 + 50/5) = 0.4 \times 11 = 4.4$$

From rearranged equation 1.5, the width of the peak at the baseline can be calculated.

$$(W_b)C = \sqrt{\frac{16(V_R C)^2}{N}} = \sqrt{\frac{16(8.4)^2}{554}} = 1.4$$

$$(W_b)D = \sqrt{\frac{16(4.4)^2}{554}} = 0.75$$

Using these data, we can establish the peak center and width at the base, as shown in Figure 1.23 by the points and the heavy lines at the baseline. We have, however, no way yet to establish the peak height for the peaks of C and D.

In order to predict the heights of the two peaks, we need more data.

Table 1.1   CALCULATIONS ON AN IDEAL LIQUID PARTITION CHROMATOGRAM[a]

		Partition Coefficient, K	Phase Ratio	Holdup Volume, V	Retention Volume, R	W_b	$W_{0.5}$	Theoretical Plates, N
Calculation of N					<u>4.0</u>		<u>0.50</u>	*554*
Prediction of Figure 1.24 for solutes C and D	C	<u>100</u>	<u>5.0</u>	<u>0.40</u>	*8.4*	*1.4*	*0.84*	*554*
	D	<u>50</u>	<u>5.0</u>	<u>0.40</u>	*4.4*	*0.75*	*0.44*	*554*

[a] Values for V, R, W_b, and $W_{0.5}$ are in mL. The underlined numbers were measured, and those in italics are calculated in our discussion.

Specifically, we need to know the relative amounts of the two solutes being separated. The values that are used to plot an actual peak come from a detector at the end of the column that measures the concentration of the solute as it emerges. As a first approximation at least, the detector readings are linear over a specific range with respect to the solute concentration, and the area traced by a given peak is proportional to the total amount of solute present. This relationship is extremely important in quantitative GC and HPLC and will be considered in more detail in Chapters 2 and 6. Again, as a first approximation the detector reponses are similar for closely related solutes. In our case, suppose that C and D are sufficiently similar and behave in an ideal fashion so that these two approximations are true. If we then separate equal amounts of C and D, the areas of the two peaks will be equal. We can then find the shapes of the two peaks by assigning an arbitrary height to one and by calculating the other. Obviously the actual peak areas will depend upon the actual amounts of the solutes present as well as the nature of the detector and the system. However, if we assign an arbitrary height of 10 to the peak of D, H_D, we can calculate the area of C, H_C, from equation 1.6, since the amounts of C and D and the areas of the peaks must be equal.

$$Area\ Triangle = \frac{1}{2}(W_b)_D H_D = \frac{1}{2}(W_b)_C H_C \qquad (1.6)$$

$$H_C = \frac{0.75 \times 10}{1.4} = 5.4$$

With relative peak heights of 10 and 5.4 for D and C, we can then erect the triangles shown in solid lines in Figure 1.23. If one wished, the $W_{0.5}$ values could be calculated or measured from the peaks in the figure. Our theoretical picture of the separation of equal amounts of two solutes, C and D, with respective partition coefficients of 100 and 50 on a column having 554 theoretical plates, a holdup volume of 0.40 mL, and a phase ratio of 5.0, is now complete.

From a strictly practical point of view, the data in Figure 1.23 show that compound D will come off the column after about 4 mL of effluent has emerged and that most of D will be in the following 0.5 mL. Solute C will elute between 8 and 9 mL. The calculations described above are usually used in the practice of chromatography, but they are of more interest to us in that they show some important qualitative ideas and relationships. Several of these concepts should be noted.

1. The efficiency of a column (either N or HETP) is independent of the solutes being separated.

2. The longer a solute stays on a column (larger V_R), the broader the band will be as it is eluted. This band or peak broadening is a function of the chromatographic or partition process and is *not a*

diffusion phenomenon.   However, the peak broadening can be reversed by temperature programming in GC and gradient elution in LC.

3. There is a direct relationship between a solute in a static situation (measured by K) and its behavior in a chromatographic system.

One more quantitative term might be of interest.   This is the peak resolution or, simply, the resolution between two solutes detected in any system.   This resolution is expressed as R (or α) and is defined in equation 1.7 for the separation of C and D.

$$R = \frac{2(V_R C - V_R D)}{(W_b)C + (W_b)D} \tag{1.7}$$

In other words, R or α is the distance between two peak maxima divided by the average base peak widths of the two solutes.   For the separation in Figure 1.23, the R value would be:

$$R = \frac{2(8.4 - 4.4)}{1.4 + 0.75} = 3.72$$

An R value greater than 1 is considered to be descriptive of a satisfactory separation.

The Equilibrium and Diffusion Problems

The data in the above sections and the quantitative relationships within it are theoretical and are based upon two ideal conditions that can never exist in a true chromatographic system.   These are *a complete equilibrium in all parts of the chromatogram* and the complete *lack of any solute diffusion.*   A complete equilibrium could exist only if there were no flow of mobile phase in the system and the system were allowed to stand as such until equilibrium was reached.   Diffusion takes place any time that solute is in contact with the system and would be zero only if it moved through the system at an infinitely fast rate.   Thus, a fast flow of mobile phase will hinder the equilibrium and help the diffusion problem and vice versa.   The observed symptoms of these problems are the spreading of spots on TLC or PC and the spreading of peaks in GC and HPLC.   Since the two aspects are in conflict with one another, one tries in a practical case to use the fastest rate that will not disturb the equilibrium too much.   The relationships between diffusion, equilibrium, and the velocity are expressed in the Van Deemter equation (equation 1.8).   In this equation

$$H = A + \frac{B}{\bar{v}} + C\bar{v} \tag{1.8}$$

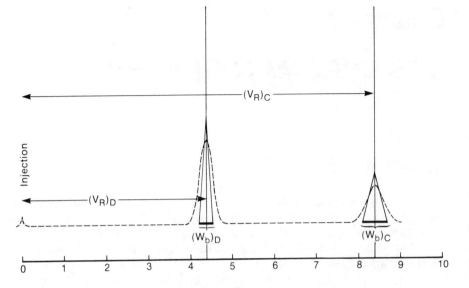

Figure 1.23 An ideal chromatographic separation of equal amounts of C + D as it might appear on a chromatographic chart. The solid lines represent theoretical peaks. The dotted lines represent the more probable peaks that would result because of diffusion.

H is the HETP and is a measure of column efficiency. Smaller HETP's mean more efficient systems. The term A is a spreading due to eddy diffusion in the column and is a function of the nature of the column. A is constant with respect to the velocity. The term B is spreading due to diffusion and is, as stated above, indirectly proportional to the velocity. The $\bar{v}$ is the average velocity of the mobile phase. The term C is an equilibrium term and is directly proportional to the velocity and peak broadening as noted above. A more quantitative treatment of these important factors is beyond the scope of this text, but can be found in most of the books noted in the Bibliography.

REFERENCES

[1]This concept is well summarized by L.S. Ettre and A. Zlatkis, Eds., in *75 Years of Chromatography-A Historical Dialogue,* Elsevier Publishing Co., New York **(1979)** pp. 285-296.

[2]L.R. Snyder and J. J. Kirkland, *Introduction to Modern Liquid Chromatography,* 2nd ed., Wiley-Interscience, New York **(1979)**.

[3]H.A. Laitinen, *Chemical Analysis,* McGraw-Hill, New York **(1960)** p. 492.

Chapter 2

GAS CHROMATOGRAPHY

2.1 INTRODUCTION

Gas chromatography, or GC, was the first of the chromatographic methods developed during the instrumental and electronic era that has revolutionized science over the last thirty years. It is now used routinely in most industrial and academic laboratories. Although the techniques and instrumentation have become complex, the basic chromatographic processes as described in Chapter 1 still prevail and, for many purposes, simple equipment is adequate. However, in recent years, an almost classic case of "more is better" has developed. The application of microprocessors to GC has simplified the operation of the ever more complex instrumentation and made the results easier to obtain, more meaningful, and easier to interpret.

The development of GC has had a most important effect on the development of liquid chromatographic (LC) methods. The application of the methods used to optimize GC, and the concepts of detection and electronic quantification of detector results, have converted classic LC methods such as column chromatography into high performance liquid chromatography or HPLC. In the last four chapters of this book various LC methods will be described, and this important evolution (or perhaps revolution) will be chronicled.

GC can be used on any mixture of compounds in which some or, preferably, all of the components have an appreciable **vapor pressure** at the temperature used for the separation. The vapor pressure or volatility allows the components to vaporize in and move with the gaseous mobile phase. In LC, the corresponding restriction is that the mixture components must have some **solubility** in the liquid mobile phase. It would appear at first glance that the vapor pressure restriction in GC is more serious than the solubility restriction in LC and, on balance, it is. However, when one realizes that temperatures as high as 400°C can be used in GC and that the chromatography is carried out rapidly to minimize decomposition, the restriction is less serious. Furthermore, in GC, non-volatile compounds can often be converted into more volatile and stable derivatives before chromatography.

GC is a rapid and precise method for the resolution of extremely complex mixtures. The time required for a separation may vary from a few seconds for simple mixtures to hours for samples containing 500-1000 components. Mixture components can be identified by their characteristic **retention times** under precise conditions. The retention time is the time that a given compound is retained on the column. It is measured from a recorder trace of the chromatogram and is directly analogous to retention volume in HPLC and *Rf* in TLC. With proper calibration, the amounts of the components of a mixture can also be accurately measured. The major disadvantage of GC is that it is not easy to use for the separation of larger quantities of mixtures. Separations on a milligram basis are easy, those on a gram basis are possible; but the separation of pound or ton lots would be difficult unless there were *no other method*. (However, it is possible, see below.)

In both GC and HPLC, columns may be reused and, with proper care, will last a long time. Such care should be taken since the columns may be very expensive. This is in contrast to thin layer or column chromatography in which the stationary phase is usually used only once.

The stationary phase in GC is normally a liquid deposited on some inert support (gas-liquid chromatography) rather than a solid functioning as a solid adsorbing surface (gas-solid chromatography). Gas-solid systems have been widely used in the purification of gases and the removal of fumes, but are less useful in chromatography. The use of liquid phases allows one to choose from a tremendous variety of stationary phases that will bring about the separation of almost any mixture. The only restriction in the choice of such liquids has been that they must be stable and non-volatile under the conditions of the chromatography. This situation is changing, however, due to the development of bonded phases and the use of the highly efficient capillary or open tubular columns. In bonded phases, the liquid is actually bonded to the solid support or capillary column wall rather than being simply deposited on it. Such materials will be briefly mentioned later in this chapter and will be discussed in Chapter 6 as well, although the HPLC column packings involve different chemistry for some of the bonded phases. With these recent developments, it is quite likely that fewer and fewer different GC liquid phases will be commonly used.

The use of detectors to continuously analyze the effluent from a chromatograph has literally made GC and HPLC possible. In GC, the varieties of detectors available, their universal application to many types of compounds, and their high degree of sensitivity have allowed the accurate determination of a wide range of components, sometimes in extremely small amounts. The availability of selective detectors, for example one that will detect only compounds containing P, N, or S has also been most important. This is in contrast to HPLC where many fewer types of less sensitive detectors are available.

In the original edition of this book, we published a sketch of a home-made GC apparatus.[1] We now believe that essentially all GC instruments are commercial, and probably should be. In 1984, the prices of systems ranged from about $3000 for a simple instrument to about $25,000 for a state-of-the-art instrument with a data processor.

The next three sections of this chapter will be devoted to operational directions for a typical GC apparatus, the choice among the variables for a GC system, and a more detailed discussion of the various components of a GC system. In the final sections of the chapter we will consider special techniques and ways to improve separations. On some topics the amount of information given is incomplete to carry out an operation, i.e., packing or coating a GC column. In this case the methods are only mentioned to provide general information.

2.2 OPERATIONAL DIRECTIONS

Even though some GC systems are quite complex, they all operate in basically the same manner. The operation will be described as a series of steps, and the subsequent section will give more information about each step. If the GC is already turned on, these directions become a series of simple checks. A commercial GC is shown in Figure 2.1.

1. The instrument is inspected, especially when it has not been in continuous use. This is done to check that the correct column is installed, that the injector septum is not defective (that is, contains a large hole from usage, which might leak), that the carrier gas fittings are tight, that the oven door is closed properly, that all the electrical parts are working properly, and that the correct detector is installed.

2. The gas flow through the column is started or adjusted. This is done by opening the main valve on the gas tank and then turning the secondary (diaphragm) valve to about 15 psi and opening the needle valve slightly. This will permit a slow flow of gas (2-5 mL/min with a packed column and about 0.5 mL/min with a capillary column) to pass through the system and protect the column and detector against oxidative destruction. In many modern instruments, the gas flow may be dialed into a rotameter or an automatic flow or pressure controller, or it may be entered via a microprocessor-based control module. Whatever the type, the fittings of the system (especially the column fittings) are checked for leaks with a soap solution, special leak detector solution (SNOOP), or a commercial leak detector. See below for more details on the gas flow.

3. The column is heated to the desired initial temperature. This is done in older instruments by turning the variable voltage transformer, which controls the heating coils in the oven, to about 90 V. When the temperature is 10-15°C below the desired temperature, the transformer is turned to a voltage (10-50 v) that will keep add-

Figure 2.1 A commercial gas chromatograph.   (Reproduced through the courtesy of IBM Instruments, Inc.)

ing enough heat to balance the heat loss.   A newer GC with a direct dial temperature control is easier to operate and does the same, but probably with less overheating.   A microprocessor-controlled GC, in which the desired temperature is entered, is the easiest to operate, and has the most accurate temperature control.   See Section 2.4 for information on choosing the column temperature, temperature program, and maximum temperature.

4. The separate heaters for the injector and detector are turned on or adjusted.   Their temperatures should be about 10-25°C higher than the final column temperature.   The detector temperature must be greater than 100°C so that water cannot condense if it is formed inadvertently or is present.   See Section 2.4 for more information.

5. The flow of carrier gas through the column is increased to 25-30 mL/min for a 3 mm (or less often 6 mm) packed column or the optimum flow rate when it is known.   The diaphragm valve should be increased to give a pressure of 60-70 psi.   This flow rate is set either as described above or by an adjust and test procedure with the column disconnected from the detector and attached to a bubble flowmeter.   If no flowmeter is available, see the section on The Carrier Gas for construction details.

6. The current to the detector is turned on only if the carrier gas is flowing to protect the filaments.. In the case of a thermal conductivity detector (TCD), the simplest to use, the current is adjusted to 150-200 mA or to the optimum amperage, when known. After the detector chamber is temperature stabilized (2-3 min), the electrical circuit is balanced so that the pen is on the strip chart recorder baseline. If the GC is equipped with a flame ionization detector (FID), the most commonly used detector, several additional checks are necessary. The FID requires hydrogen for the flame, and thus the hydrogen generator must be turned on and the flow adjusted to be the same as the column flow (25-30 mL/min). The air (oxygen) for the detector is started and it is regulated to about ten times the column flow. (The optimum flow for any system can and should be determined by testing.) The flame in the FID can then be ignited by pressing the igniter button on the GC. A slight pop will indicate when the flame ignites. Stabilization generally occurs in 2-3 min. The detector circuit is balanced to place the recorder pen to the baseline of the recorder chart.

7. The sample is injected. A small amount (see below) of the liquid (careful, no **overloading**) or a solution of sample in a volatile solvent plus a small amount of air with a TCD (to give an **air peak** or to mark the zero time) is taken up into a microsyringe equipped with a long needle. The FID will sometimes give a zero time peak due to a slight flow change when the sample is injected. The sample is placed on the column by carefully forcing the needle through the rubber injection port septum to its fullest extent and immediately forcing it out of the syringe as fast as possible. The syringe should

Table 2.1   SAMPLE SIZE AND DETECTOR TYPE

Normal Sample Size	Detector
10-100 μL	TCD-Normal
1-10 μL	TCD-Small volume
1-10 μL	FID
0.1-5 μL	ECD

TCD – thermal conductivity detector
FID – flame ionization detector
ECD – electron capture detector

then be removed rapidly and cleaned with solvent. A GC equipped with a normal TCD requires at least 10 μL of sample and an FID about 1-5 μL. Table 2.1 gives more information on amounts. A capillary GC column requires special injection techniques (splitting) to lower the injected amount to less than 1 μL. Section 2.4 gives more details on splitting and amounts.

8. The peaks are recorded to produce the chromatogram. This is done on either a strip chart recorder or some type of data system that gives a printout and plot after the run is complete.

Figure 2.2 shows the chromatograms that resulted when mixtures of

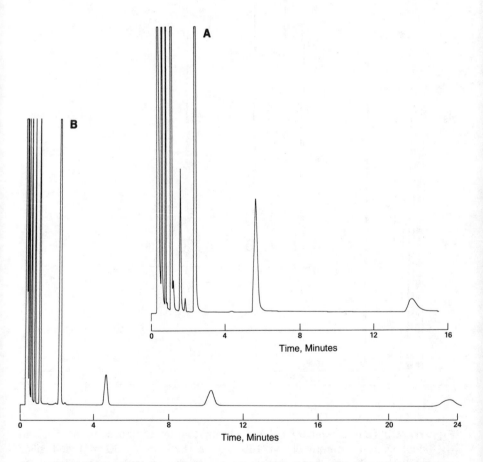

Figure 2.2 (a) A packed column chromatogram of a sample of C_8 to C_{14} alcohols run isothermally at 150°C. The column was glass, 3 mm x 2 m, and the stationary phase was methyl silicone oil (SE-30) on Chromosorb W. **(b)** A packed column chromatogram of a sample of C_5 to C_{22} hydrocarbons run isothermally at 170°C on the same column as Figure 2.2a.

C-8 to C-l4 straight-chain alcohols and C-5 to C-22 straight-chain hydrocarbons were separated in a 3 mm x 2 m glass column that was packed with
3% methyl silicone oil (SE-30) coated on 80/100 mesh Chromosorb W HP
at a constant temperature (isothermal mode) with nitrogen as the carrier
gas. Figure 2.3 shows the same mixtures separated on a 0.25 mm by 30 m
fused silica capillary column coated with 1 μm of the same stationary phase
and using the same carrier gas and GC equipment. Different temperatures
were required for the separations because of the differences in boiling

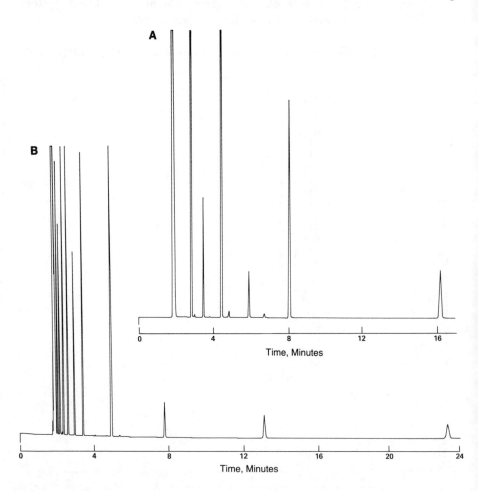

Figure 2.3 (a) A capillary column chromatogram of the sample of C_8 to
C_{14} alcohols run isothermally at 180°C in a 0.25 mm x 30 m fused silica
column with a 1 μm phase thickness. Note the narrower peaks from the
capillary column as compared to those from the packed column in Figure 2.2a.
(b) A capillary column chromatogram of the sample of C_5 to C_{22} hydrocarbons run on the same column as Figure 2.3a. This chromatogram should
be compared to that in Figure 2.2b.

points of the samples and because of the differences in overall efficiencies of the packed and capillary columns. The high efficiency and quality of both columns as shown by the **baseline separation** and symmetrical peak shapes is obvious. The enhanced efficiency of the capillary column is indicated by the narrower peaks. All four chromatograms shown in Figures 2.2 and 2.3 were recorded on a microprocessor-based data system, and an FID detector was used. The printouts from the capillary column separations of the alcohols and hydrocarbons are shown in Figure 2.4, and they clearly indicate the retention times and peak area percents. The contrast in ease of data analysis between the computer-generated results and a strip chart recorder and manually done retention time and areas is also obvious. The retention times will give information useful for identifying the compounds, and the areas under the peaks will give the required quantitative information. These chromatograms can be used to illustrate a number of critical concepts and features of GC.

The separation power of the capillary column for a sample with many components, such as lemon oil, as compared to the same separation with a packed column is shown by the two chromatograms in Figure 2.5. In this case the capillary column had a plate count of 3400 plates/m (3400 x 15 = 51,000 total plates) and the packed column was 15,200 plates/m (15,200 x 6 = 91,200 total plates). In comparison, the best HPLC columns have in excess of 100,000 plates/m.

RT	AREA	TYPE	AREA %	RT	AREA	TYPE	AREA %
2.35	20.53	BV	0.005	1.70	25.78	BB	0.005
2.40	61.66	VV	0.015	1.97	119.65	PB	0.025
2.54	406686.00	VV	98.140	2.26	22.83	BV	0.005
2.63	397.91	VV	0.096	2.32	69.97	VV	0.015
2.67	283.64	VV	0.068	2.45	464568.00	VV	98.250
2.76	565.11	VV	0.136	2.66	809.87	VV	0.171
2.94	26.00	VV	0.006	2.84	34.51	VV	0.007
3.03	17.34	VB	0.004	2.92	20.33	VB	0.004
6.58	2265.89	BB	0.547	3.21	694.10	BB	0.147
8.12	236.68	BB	0.057	4.30	629.25	PB	0.133
9.70	2397.09	BB	0.578	5.65	480.52	BB	0.102
10.15	16.45	BB	0.004	7.15	763.43	BB	0.161
11.18	131.21	BB	0.032	8.68	1095.93	BB	0.232
11.81	14.35	BB	0.003	10.18	478.57	BB	0.101
12.64	820.69	BB	0.198	11.64	821.03	BB	0.174
15.31	350.19	BB	0.085	14.34	1515.73	BB	0.321
17.95	101.98	BB	0.025	16.77	217.41	BB	0.046
				18.99	232.05	BB	0.049
				21.51	241.58	BB	0.051

TOTAL AREA = 414393.00
MULTIPLIER = 1

TOTAL AREA = 472840.00
MULTIPLIER = 1

Figure 2.4 Integration output on the capillary column chromatograms in Figure 2.3 That on the left is from the alcohols and on the right from the hydrocarbons. RT is the retention time.

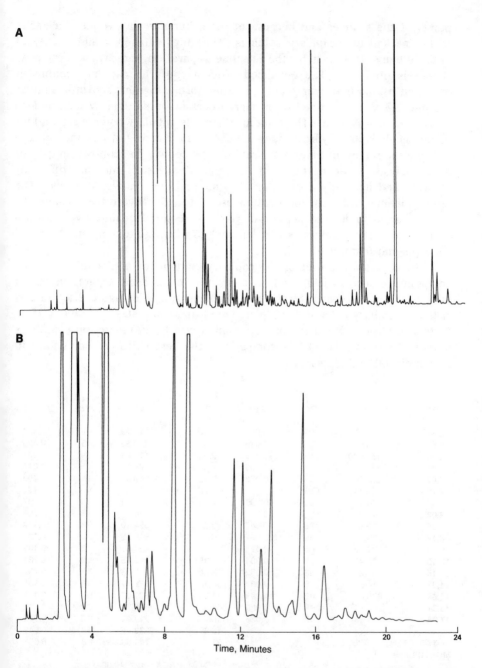

Figure 2.5 (a) A capillary column separation of a sample of lemon oil. The column was 0.25 mm x 15 m coated with methyl silicone oil (SE-30) and temperature programmed from 100-200°C at 5°/min. **(b)** A packed column separation of lemon oil. The column was that used in Figure 2.2 and was temperature programmed from 70-170°C at 5°/min. Note the broader peaks from the packed column.

2.3 CHOOSING A SYSTEM

There are four major variables in GC. These are, in order of increasing complexity, the carrier gas, the type of detector, the type of column and stationary phase, and the temperature or temperature conditions for the separation. These variables will be discussed in the same order. It is interesting to note the differences between GC and HPLC, for in HPLC the detector is the easiest choice, followed by the temperature (usually at room temperature) and the column, with the mobile phase liquid choice most difficult.

The Carrier Gas

The factors that cause a substance to be transported through a GC column are the volatility inherent in the substance itself and the flow of the gas through the column. The gas flow is described by two variables, the flow as measured in mL/min and the **pressure drop** between the beginning and end of the column. The exact nature of the carrier gas is usually secondary as far as separations are concerned, but a small effect on **resolution** is possible, as discussed in the next section. The choice of the carrier gas depends to some extent on the detector used: thermal conductivity, flame ionization, electron capture, or element specific.

Nitrogen, helium, argon, hydrogen, and carbon dioxide are most often used as carrier gases, for they are unreactive and can be purchased pure and dry in large-volume, high-pressure tanks. It is critical that the purest available gas be used to lessen **detector noise.** In most cases, the gas should be dried even more thoroughly with a molecular sieve drying tube, and oxygen should be removed with an oxygen trap. Fortunately, either of these two adsorbents, which are often in the same cartridge, will also trap any oil coming from the gas tank. When capillary GC is to be carried out, the requirements for gas purity are even more stringent. In this case it is critical that even the valve on the gas tank and the flow controller be checked to make sure that a stainless steel diaphragm is being used rather than a polymeric one. The latter will **bleed** sufficiently to give additional detector noise (background).

Although helium or hydrogen will give the greatest sensitivity to a TCD (the conductivity is dependent on the mass of the gas), they are slightly inferior to nitrogen in that more lateral flow and mixing occurs with the less dense gases. Despite this minor disadvantage, helium is generally used. An instrument is normally set up with only one carrier gas, and it is rarely necessary to change. Certain ionization detectors require argon, that is considerably denser and has a correspondingly slower flow (greater pressure drop). Nitrogen is generally used with a FID, although once again, any of the gases can be used. It will be shown in the next section that the FID response and, thus, its ultimate sensitivity is slightly affected by the choice

Table 2.2   CARRIER GASES AND DETECTOR USAGE

Carrier Gas	Detector
Hydrogen	Thermal conductivity Photo ionization
Helium	Thermal conductivity Flame ionization Photo ionization Flame photometric Thermionic Hall electrolytic
Nitrogen	Flame ionization Electron capture (DC mode) Photo ionization Flame photometric Thermionic Hall electrolytic
Argon	Flame ionization
Argon + 5% methane	Electron capture (pulse mode)
Carbon dioxide	Thermal conductivity Photo ionization

of the mobile phase. An electron capture detector (ECD) for the halogens requires either nitrogen or argon plus methane (5-10%). The element specific detectors for S, P, and N require either helium or nitrogen.

Capillary columns with their very low flow rates, 0.1-2 mL/min, use nitrogen, helium, and hydrogen. The flow rate should be adjusted to give the maximum detector performance. A TCD can be used with capillary columns if it is sensitive enough. These suggestions for carrier gases are summarized in Table 2.2.

Compressed air, although it is readily available, cannot be used because oxygen will oxidize the stationary phase, the detector, and the compounds undergoing separation. However, in simple GC with a TCD, it is possible to use natural gas (propane or butane from the laboratory gas jet).

The Detector

The first choice of a detector in GC is now the FID. It used to be the TCD because it was simple, stable, and versatile. However, if the GC is

Table 2.3   SENSITIVITIES OF MAJOR GC DETECTORS

Detector	Sensitivity, g
Thermal conductivity	
Standard	$10^{-6} - 10^{-7}$
Micro	$10^{-8} - 10^{-9}$
Flame ionization	10^{-10}
Electron capture	$10^{-12} - 10^{-13}$
Photo ionization	10^{-12}
Flame photometric	$10^{-11} - 10^{-12}$
S, P compounds	
Thermionic	$10^{-12} - 10^{-13}$
P, S, N compounds	
Hall electrolytic	10^{-10}
N, halides	

not equipped with a TCD, or if the installed TCD is not sensitive enough for **trace analysis** or capillary GC, the FID (a hydrogen/oxygen flame) should be used because of its high sensitivity to every kind of organic compound.   One important feature of a TCD is that it does not destroy the compounds being detected, allowing them to be isolated or trapped.   This **sample trapping** is used for preparative GC.   Our recommendation of the FID includes a special feature, for it does not detect a common solvent for GC, carbon disulfide.   Information on the optimization and operation of these detectors is given in the next section.

When the measured sensitivities of the major detectors are examined (see Table 2.3), it is noted that the ECD, used mainly for chlorine-containing compounds, is more sensitive than the FID for these substances. The photoionization detector, PID, is about as sensitive as the FID.   The element specific detectors for N, X, P, and S are used for the **selective detection** of certain types of compounds in mixtures.

The Stationary Liquid Phase

Two aspects of a stationary phase must be considered. These are, first, the way in that the liquid is held in the column and, second, the actual chemical nature of the liquid. Traditionally, the liquid was held on the surface of some finely divided solid (such as diatomaceous earth) in the column. Examples of separations on such media are given in Figures 2.2a and 2.2b. The stationary liquid can also be deposited on the walls of a long fine capillary such as the one (30 m long and with an inside diameter of 0.25 mm) used for the separations shown in Figures 2.3a and 2.3b.

Several hundred liquid phases have been investigated in GC, and it may appear that the choice of one of them for a given separation is almost hopeless. However, relatively few phases are in common use, and they fall into certain categories, appreciably simplifying the choice. In Table 2.4, some of the phases that have been successfully used for specific separations are given. Many more can be found in the chemical literature and in the books cited in the Bibliography in the back of the book..

The principal characteristic required for a stationary liquid phase is that it will dissolve the materials being separated to a greater or lesser extent. Since the materials being separated must also have some volatility, they will then partition (Chapter 1) themselves between the liquid and gas phases. The major concept guiding the choice of a liquid phase is: *"like dissolves like."* Thus, a liquid phase should have functional groups similar to those in the sample. For example, relatively non-functional and non-polar compounds such as hydrocarbons, ethers, and alkyl halides are best separated on high molecular weight hydrocarbon liquids such as Apiezon grease. At the other extreme, such polar compounds as alcohols and amines are best separated on Carbowax, a high molecular weight polyethylene glycol.

Many samples, however, consist of materials of widely varying polarities. For these separations, liquid phases of medium polarity and multiple functionality can be used. For example, nonyl phthalate has long hydrocarbon chains (the nonyl groups) for less polar materials, carbonyl groups for oxygenated compounds, and aromatic rings for aromatic compounds. In fact, there is a wide overlap between the separating abilities of the various phases. There are also ways to calculate and predict ideal stationary phases, for example, using McReynolds constants and the Window Diagram Procedure.

Some liquid phases contain inorganic ions that can complex with components in the sample in a reversible fashion. The classic example of this technique is the separation of alkenes on liquid phases containing silver ion. The extent of complex formation between Ag^+ and a double bond is quite sensitive to the nature and stereochemistry *(cis* or *trans)* of the alkene, and excellent separations of quite similar compounds can be made. The technique is also used in TLC and other forms of liquid chromatography.

At present, the use of bonded phases in that the liquid is covalently bound to the support appears to be more important in HPLC than GC.

Table 2.4   STATIONARY PHASE CHOICES FOR
              SEPARATION OF COMMON MIXTURES

Compound Type	Stationary Phase[a]
Gases	Molecular sieves, porous packing
Aliphatic hydrocarbons, fuels	Apiezon L, methyl silicone oils
Aromatic hydrocarbons	Phenyl/methyl silicones
Alcohols, polyols	Carbowaxes (polyethylene glycols)
Carbohydrates	Cyanosilicones
Acids	Carbowaxes
Fatty acid esters	Cyanosilicones
Aldehydes, ketones	Alkyl phthalates
Esters, polyethers	Methyl silicone gums
Phenols	Methyl silicone oils
Amines	Carbowax + KOH
Drugs	Phenyl/methyl silicones
Pesticides	Phenyl/methyl silicones
Steroids	Methyl silicone gums, Dexsils
High boiling compounds	Dexsils (polycarboranylene siloxanes)
Mixtures	Carbowax esters

[a] *The fused silica capillaries currently have methyl silicone gum, methyl/phenylsilicone gums, cyanosilicone, and polyethylene glycol (plus KOH modified for amines) as the standard stationary phases. These phases are either as a liquid film or the more stable bonded film. These four phases, from non-polar to highly polar, are capable of separating almost every mixture.*

However, the fact that these bound phases do not bleed gives them a distinct advantage when the GC is connected to other types of instruments; see Section 2.5.

   In summary, a wide range of phases is available, but it is probable that a rather small number can be used for the separation of most samples.

2.4 THE SYSTEM

The Column Temperature

GC is based upon two properties of the sample being separated, its solubility in a given liquid and its vapor pressure or volatility. Since vapor pressure is directly dependent on temperature, it follows that the temperature is a major factor in GC. The separation can be carried out at a constant temperature, that is, **isothermally,** or with a variable-controlled or **programmed temperature** change.

Although column temperatures may range from -100°C to as high as 400°C, several practical limitations must be considered. Some stationary phases are solids at lower temperatures; for example, Carbowaxes are solids below 50°C and some silicones (methyl silicone gums) are solids below 100°C. Furthermore, the maximum temperatures for use with the various phases are dictated by their stabilities. At higher temperatures, a column phase may bleed or slowly decompose, giving off small fragments that produce a high background in the detector trace. At even higher temperatures, the phase may decompose completely. The minimum and maximum temperatures for the phases recommended in Table 2.4 are given in Table 2.5. A typical range for most columns is 50-300°C.

In general, better separations are obtained at lower temperatures. This allows the liquid phase to function more effectively as a dissolving medium, and small differences in solubility properties can be translated into better separations. If the temperature is too high, all of the components are in the vapor state; solubility in the stationary phase is low; and no separation results. A useful rule-of-thumb is that a 30°C rise in temperature will halve the retention time.

Isothermal GC. Isothermal chromatography is best used in routine analyses or when a fair amount is known about the materials being separated. A good initial choice is a temperature a few degrees below the boiling point of the major mixture component. There are, however, several problems with isothermal separations.

The first, of course, is the temperature choice. If it is too high, the components will elute without separation. If it is too low, high boiling components will come off very slowly or may even remain on the column to sabotage later chromatography. If the latter is suspected, the column can often be cleaned by **back flushing,** that is, by passing a stream of carrier gas back through the heated column for some time.

The second problem in isothermal work is inherent in the chromatographic process and is presented as a problem here only because it can be solved by a programmed temperature system. The longer a given sample stays on a column, the broader will be the peak. This is explained in Chapter 1 and is shown graphically in Figures 2.2a and 2.2b. Peak broadening is unavoidable. However, when the temperature is raised during the

Table 2.5   MINIMUM AND MAXIMUM TEMPERATURES
FOR STANDARD GC STATIONARY PHASES

Stationary Phase	Minimum Temp., °C	Maximum Temp., °C[a]
Apiezon L	50	225
Methyl silicones	0 (Gum, 100)	300 – 350
Phenyl/methyl silicones	0	300
Carbowaxes (polyethylene glycols)	10 – 30	225
Cyanosilicones	0	275
Alkyl phthalates	20	225
Dexsils (polycarboranylene siloxanes)	50	450

[a] *The fused silica capillary columns have a 350°C maximum for all the bonded phases and a 250°C maximum for the liquid Carbowax phase.*

chromatography, as in a programmed run, the higher boiling materials are pushed off the column more quickly with subsequently less broadening (see Figures 2.6a and 2.6b). This is sometimes called *peak compression* and can be done in column chromatography and HPLC by gradient elution (Chapters 5 and 6).

Programmed Temperature GC. In this type of GC, the temperature is raised from some given point to another at a controlled and known rate over a given time. The process can be carried out in an infinite number of ways. The increase can be (1) linear at a chosen rate, (2) stepwise, (3) isothermal followed by a linear increase, (4) linear followed by an isothermal time, or (5) multi-linear (different rates for different times). These are shown in Figure 2.7. These temperature changes can be carried out manually, but are most conveniently done with an electronic or microprocessor-controlled system.

Temperature programming is ideal for exploring the properties of a completely unknown sample and can be used to predict a good temperature for an isothermal separation. For example, if the unknown is chromatographed between 50 and 250°C (depending on the liquid phase) at about

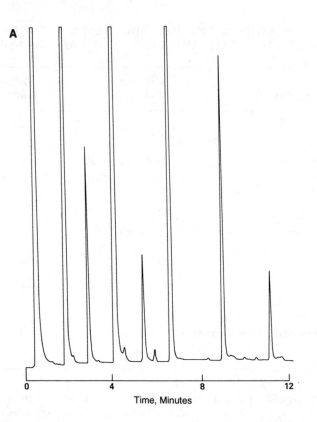

Figure 2.6 (a) A temperature programmed, 100-200°C at 10°/min, chromatogram of the alcohol sample separated isothermally in Figure 2.2a. Note the better spacing of the peaks.

8° per min, one can obtain the approximate boiling points of the components and can ascertain that there are no high boiling materials. After such an initial experiment, one can often establish a closer range for a given separation, generally from a few degrees below the boiling points of the components to a few degrees above, with a slower temperature increase, perhaps 1-2° per min. Temperatures is such a range would also be suitable for isothermal work.

There are some problems with programmed systems. Precise temperature control is hard to obtain on older instruments without microprocessors, precision heaters, well-designed air circulation, and carefully insulated ovens. Furthermore, increased temperatures will change the gas flow. All of these factors make the chromatography less precise and retention times harder to duplicate.

The most serious problem, however, with programmed systems is the effect of the higher temperature on the detectors. This is less important with an FID that operates with a high-temperature hydrogen flame, but can

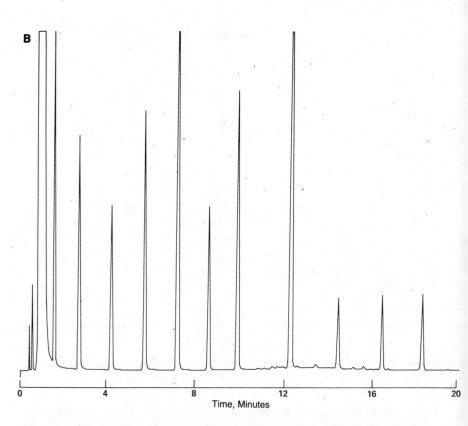

Figure 2.6 **(b)** Temperature programmed, 100-200°C at 10°/min, chromatogram of the hydrocarbon sample separated isothermally on the packed column in Figure 2.2b.

be quite serious with a TCD, which operates best at a constant temperature. This problem is partly overcome in all commercial instruments which provide a reference gas stream heated to the same temperature as the sample stream.

The Oven. Temperature control in GC is brought about by placing the column in an oven. Such an oven must be carefully insulated with a tightly fitting door. The oven should be designed such that heaters can produce a uniform and accurately sensed temperature. Furthermore, the oven should have a low mass so that it can be heated and cooled rapidly.

The Carrier Gas

Carrier gases or gaseous mobile phases have been discussed in a previous section, and their use with the various detectors is summarized in Table 2.2.

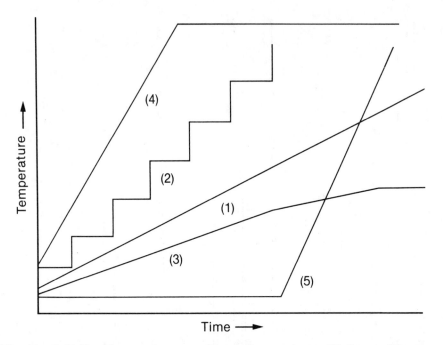

Figure 2.7 Five types of temperature programming: (1) linear, (2) step-wise, (3) multi-linear, (4) linear-isothermal, and (5) isothermal-linear.

For any given separation, there is an optimum flow rate for the carrier gas. This will depend, first of all, on the diameter of the column. The approximate ranges are 50-70 mL/min for a 6 mm (i.d.) column to 25-30 mL/min for a 3 mm column to 0.2-2 mL/min for a capillary column. Strictly speaking, since the flow is directly related to the cross section of the column, and the cross section depends on the *square* of the radius (area of a circle = πr^2), a doubling of the diameter would require a flow rate four times that of the smaller column. Thus, if a good separation is obtained on a 2 mm column at 20 mL/min, a 4 mm column would require 80 mL/min to give a comparable result. It is obvious that a smaller column can produce an appeciable saving in carrier gas.

The flow rate can be conveniently measured with a bubble flowmeter if the instrument is not equipped with a device for the purpose. A bubble flowmeter can be made from a 10 mL graduated pipet and soap solution. The pipet is dipped into the solution (to get a bubble) and connected to the outlet tube from the GC. The rate of travel through the pipet is measured with a stopwatch.

Within the wide ranges cited above, there is an optimum flow rate for any given column system. This can be determined in the following manner. Identical amounts of the sample are chromatographed at different flow rates, and the flow rate that produces the maximum peak heights is the best.

The reasoning behind this simple exercise is rather tortuous. At the optimum flow rate, the minimum amount of band broadening will occur and the system will have a maximum number of theoretical plates (these concepts are defined and interrelated in Chapter 1). The amount of band broadening can be read from the recorder chart for a given peak, but it is usually small and hard to measure accurately. Since the area under the peak is a constant with identical sample size, a lessening of band broadening *must* be accompanied by an increase in peak height, that is easy to see and measure. Thus, peak height is a measure of efficiency when identical samples are chromatographed under different conditions.

The determination of an optimum flow rate when using a flame ionization detector is more complex because one must actually deal with three flow rates: the carrier gas, the hydrogen flow rate to the detector, and the air flow to the detector. The hydrogen flow should be optimized first as described for the carrier gas. The air flow but is less crucial, is about ten times the hydrogen flow, and usually is about the same as the carrier gas flow. The carrier gas flow can then be adjusted to give maximum peak height. It may be necessary to repeat the sequence for best results. As a practical note, when the carrier gas flow and the hydrogen flow are not correctly balanced against one another, the hydrogen flame will be hard to ignite and keep going. A noisy flame is shown by a high and erratic background on the recorder chart. The presence of water in one of the gases will also give a noisy flame.

Capillary columns are operated at very slow rates, 0.2-2 mL/min, and are usually controlled at a constant inlet pressure rather than at constant flow rate. In such a constant pressure mode, the velocity of the carrier gas will increase with an increase in temperature (temperature programming). The newer and microprocessor controlled systems correct the flow changes with temperature. It may be necessary to try several gas pressures to see which one gives a maximum peak height (or maximum efficiency). With the very low flow rates in capillary systems, a **make-up gas** is normally used with most of the detectors. Such a gas is added to the effluent after it leaves the column and before it reaches the detector. It is generally the same gas as that used for the chromatogram, but is most often helium.

All of the above remarks apply when one is using a given carrier gas and attempting to devise the most efficient separation conditions. Complications arise when one may wish to change gases, since the various gases operate most efficiently at different flow rates as shown in Figure 2.8. Thus, nitrogen is most efficient, at about 10 mL/sec (0.17 mL/min) whereas helium is more efficient at 40 mL/sec (0.67 mL/min). The reason for this difference, as noted above, is that samples diffuse more rapidly in gases of lower molecular weight. Thus, the chromatography must be carried out at a high flow rate to minimize diffusion with the less dense gases. The choice often becomes one of cost, for tanks of pure nitrogen are much cheaper than tanks of helium or hydrogen.

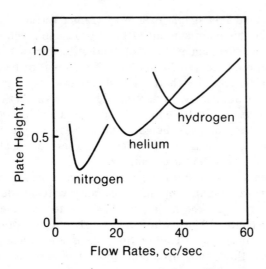

Figure 2.8 Plots of the change in efficiency (in plate height) vs carrier flow rates for the three standard carrier gases in a capillary column.

In recent years, and mainly due to more controllable systems, **flow programming** has become more common. In such systems, the flowrate is changed with time. Such programs may be used to enhance the separation of certain portions of a mixture while decreasing the overall time required for a separation.

The Injection Block, Injection Port and Splitter

The sample to be chromatographed is introduced with a syringe into the **injection block** through the **injection port,** normally a hole covered by a rubber septum. A schematic is shown in Figure 2.9. The block must be heated independently from the column, and is normally held at a temperature 10-15°C higher than the maximum column temperature. Thus, the entire sample is vaporized as soon as it is injected and is swept into the column. There are two main problems with injectors. The first is sample decomposition and the second is any residues left after incomplete volatilization of the sample.

Decomposition may occur if the block is at too high a temperature, if there are hot spots in the block due to uneven heating, or if there is any metal exposed to the sample. If too high a temperature or a hot spot is suspected, chromatograms should be carried out at lower injector temperatures and the results compared. Decomposition will be evidenced by extraneous peaks or extra shoulders on peaks at the higher temperature. Modern injector parts in contact with the sample are usually glass or glass-lined

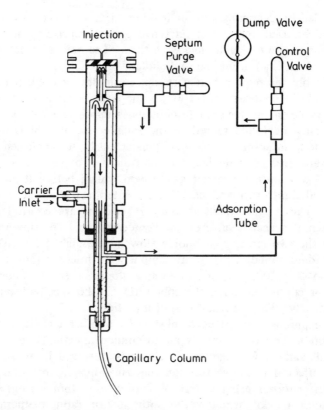

Figure 2.9 Schematic of a GC injector. (Reproduced through the courtesy of Scientific Glass Engineering, Inc.)

metal, and hot spots are rarely seen. Such glass or glass liners are silanized (see below) to prevent sample adsorption.

The injector may become dirty, either due to incomplete sample volatilization, or decomposition, or both. In this case, **ghost peaks** may be seen in subsequent chromatograms as these residues slowly volatilize, often due to their further decomposition, or because the residues adsorbed sample in subsequent runs and portions of the sample are slowly desorbing. The injector can be cleaned by scraping away the residue, washing with solvent, or with hot water, or with any combination thereof. It is also possible to inject a large sample of water and/or solvent to clean the injector *with the column disconnected.* Glass liners or inserts can be replaced. In injector design it is important that all the parts of the injector be as small as possible to lessen the **dead volume** of the injector, and that no **unswept volume** (a space where the gas flow is held up) results.

The septum (Figure 2.9) is normally made of silicone rubber faced with Teflon or some heat stable polymer. The major problem with septa is

bleeding, caused by decomposition or leaking. For very accurate work, a septum should be used only for about ten chromatograms, and an attempt should be made to use the same hole each time. The newest injectors have needle guides to facilitate using the same hole.

Capillary columns require that very small samples be used, perhaps as low as 0.01 μL in contrast to 1-100 μL in packed columns. Since the accurate measurement of such small amounts is impossible with a syringe, some means must be found to reduce the sample size after it is injected. This can be done in several ways, but it most easily accomplished with a **stream splitter.** In such a device, a known amount of sample is injected in the normal way into the carrier gas stream which, before it enters the column, is divided into two streams. One is vented, and the other goes into the column. The relative flows in the two streams are controlled by a restriction such as a needle valve on the vented stream. The flow rates are measured on the two streams (a bubble flow meter might be used) and a **split ratio** is determined. If a 1 μL sample were injected into a splitter having a ratio of 1:100, 0.01 μL of sample would enter the column. In microprocessor controlled units, the split ratio can be specified and produced automatically. The net result is **split injection.**

The same purpose can be accomplished by another technique called a splitless injection, or more correctly an on-column injection. In this technique, a small and known amount of sample is dissolved in a low boiling solvent and injected (rapidly) through the usual heated injection system onto a *cooled column* using a syringe with a long thin syringe needle. When the oven is quickly heated to the isothermal operating temperature or, more commonly, through the temperature program, the low boiling solvent is swept from the sample by carrier gas, leaving the sample deposited in the first part of the column. The heating program will volatilize the sample and allow separation to take place. This mode of injection gives a **solvent effect,** that leads to peak sharpening.

Most injection systems are commercial, and a detailed discussion of any but the simplest, as shown in Figure 2.9, is beyond the scope of this text. Books listed in the Bibliography have considerable information.

Sample Preparation and Injection

The ideal sample for GC should contain only those materials that will be separated on the column plus, in most cases, a volatile solvent. Although both liquids (not in solution) and even volatile solids may be injected directly, most samples are used in solution in pure, dry organic solvents. Concentrations generally range from 1 to 10%. Non-volatile materials or those much less volatile than the sample should not be present, since they will be deposited in the injector of column and destroy their effectiveness. The most common solvents are the low boiling hydrocarbons, ethyl ether, alcohols, and ketones.   When a solvent is used, it will, of course, show up on the recorder trace and be discounted.   Obviously, solvents should be selected that will have appreciably different chromatographic properties from the sample.   The use of carbon disulfide with an FID is rather special, for this detector is insensitive to this solvent and one can ignore the solvent peak.   However, carbon disulfide has disadvantages in its poisonous nature and its extremely low ignition temperature.

The introduction of gases can be carried out with a gas-tight syringe or an especially designed gas inlet system.

The preparation of suitable samples for GC will depend upon the source and value of the crude material and may involve extraction, preliminary distillation, and the usual techniques of organic chemistry.

Various types of materials are sometimes added to GC samples for specific purposes.   The most common of these is air or sometimes methane. These gases are not retained at all by most columns, and the time they take to move through the system is taken as a measure of the volume of the mobile phase.   The resulting peak is called a **zero time** peak.   The solvent is also used to give a peak from which retention times can be measured.

Known compounds are frequently introduced into GC samples for such purposes as the identification of a suspected component (the **spiking** technique) or as an internal standard for quantitative work.   These will be discussed later.

Successful GC requires that the samples have some volatility for separation.   Frequently, non-volatile samples are converted to volatile derivatives that are then chromatographed.   A common example of this technique is the conversion of the non-volatile carbohydrates into their trimethylsilyl ethers, which are chromatographed quite easily.   Samples may also be converted into derivatives that have elements that show up in element specific detectors such as the nitrogen or phosphorus detector.   The use of such reactions in the heated injector is called **reaction gas chromatography.**

Most sample injection in GC is done with microsyringes, and pictures of several of these are shown in Figure 2.10.   In the figure, note the metal guides on side of the one syringe.   These can help to ensure that the plunger goes straight into the syringe body without bending and to achieve repeatable sample injection.   Automatic sample injectors and microprocessor controlled systems are available, but beyond the scope of this discussion.

The critical requirement in sample injection is that the proper amount of sample be placed in the right place in the shortest possible time.   This may be into a heated injector block or directly onto a column as described in the previous section.   If the sample is not injected properly, such as from a syringe only partly in the injection port, broad tailing peaks may result, as shown in Figure 2.11.   Sometimes, an ingenious technique known as **solvent flush** is used to make sure that all of the sample gets from the syringe into the system.   A portion of pure solvent is drawn into the syringe needle *before* the sample.   Thus, the sample is followed into the system by a plug

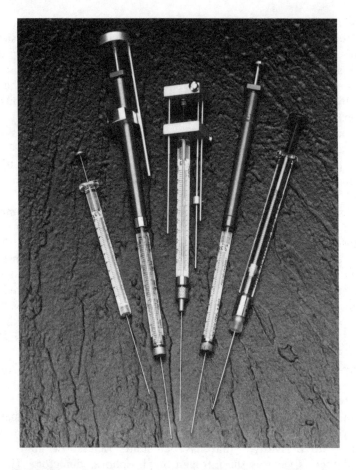

Figure 2.10 Pictures of various types of syringes.   The one with the Chaney adapter indicates one way of protecting the fragile plunger. (Reproduced through the courtesy of the Hamilton Co.)

of pure solvent to clean the syringe. If air or methane are added to the sample, they are normally pulled into the syringe after the sample.

The injector system may be tested in two general ways to show that it is operating satisfactorily. Samples containing known ratios of two compounds may be injected. If the peaks do not come out in the same ratio, **sample discrimination,** or selective treatment of one of the components, is taking place. This could be due to unequal volatilization because of too low a temperature, or may be due to selective adsorption on the injector walls or residues on the walls, or many other things. The injector system can also be checked by heating it with the normal stream of carrier gas, but without injecting a sample. If peaks appear, they are derived from sample residues in the septum or in the injector.

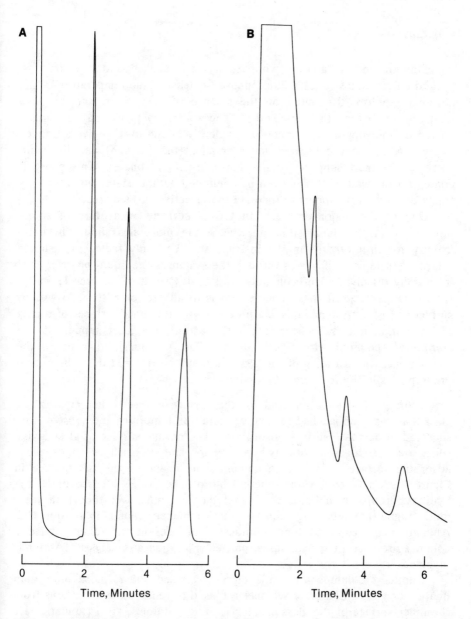

Figure 2.11 (a) A chromatogram of C_{14} to C_{16} hydrocarbons in isooctane, properly done. **(b)** A chromatogram of the same sample showing a broad solvent peak, broadened hydrocarbon peaks, and tailing peaks because the syringe needle was not inserted far enough into the injection port. In this case the sample backflashed to adsorb on the septum.

Columns

Columns in GC are of two basic types: packed and capillary.   A packed column consists of a liquid phase (at least at the temperature of the chromatography) distributed on the surface of some inert support in a relatively large tube (1-3 mm, i.d.).   The stationary phase may be simply coated on the support, or it may be covalently bound to it to give a bonded phase.   A capillary column is much smaller (0.2-0.2 mm) and the walls serve as the inert support for the stationary liquid phase.   This phase is coated on the walls and may even be combined with a small amount of very finely divided inert support to increase the effective surface area.

One of the major problems in GC is **surface adsorption** of solute molecules.   In an ideal situation, the solute molecules should have no affinity for *any surface* in the instrument.   The only factor that should retard the sample as it passes through the system is the liquid phase in that it partially dissolves.   Unfortunately, and as described in Chapter 1, one of the fundamental properties of molecules is to adhere, or adsorb, to various surfaces.   In GC systems, this is a problem with the inner surfaces of tubing and the injection and when, to a large extent, the inert support is not *completely* covered with liquid phase.   The problem shows up in the chromatogram as a **tailing** of the peaks and will be referred to in the appropriate places in the following discussion.

Tubing.   The ideal material for the surface is fused silica (quartz) or glass that has been leached to remove metal ion impurities and subsequently silanized to decrease surface adsorption.   The normal surface of glass is like silica, that is, covered with hydroxy groups and, as such, is an excellent adsorbing surface.   For use in columns, the inside of the tubing (see in Figure 2.9) is reacted with trimethylchlorosilane [$(CH_3)_3Si$] to block the hydroxyl groups as silyl ethers.   Coated glass columns, even though they are quite fragile, are replacing stainless steel or nickel columns in routine GC.   Aluminum and copper tubing is undesirable.   Teflon and other polymeric columns are used in special cases, particularly when an analysis of water is involved.

Capillary columns are made either of fused silica (more pure than quartz) or glass.   Elaborate schemes are used to remove the metal ions from the inner surface of the glass capillary tubing.   Stainless steel capillaries are rarely used because of their adsorption as compared to the fused silica or glass.

Size and Length of Columns

The length of a GC column, or for that matter, the length of any chromatographic system, is a matter of compromise.   The peak broadening that is inherent in the chromatographic process, as well as that from diffusion and other problems, increases with the length of time the sample is in the system. Thus, two compounds that have very similar properties may be theoretically separable on a very long column. However, the separation may be destroyed by the band broadening due to the length of the column, although, as noted in Chapter 1, band broadening is greatly reduced in more efficient systems. In the early days of GC, relatively long packed columns of 4-6 m or more were used. As better and more efficient packings have been devised (narrower distribution of particle sizes as well as more inert), the length has decreased to about 2 m. Capillary columns, in contrast, started out at 100 m and are currently 15-25 m. In practice, the length is governed by the pressure drop of the column. If the mesh size of the particles is too small, and if the particle size distribution is too broad, it may not be possible to push the carrier gas through a long packed column.

The tubing, whether metal, glass, or plastic, is generally coiled so that it will fit into a fairly compact oven. Some data on column sizes and sample sizes are given in Table 2.6.

Supports. The purpose of a support in GC is to provide an inert, non-adsorbing surface onto which the stationary liquid phase may be deposited. Such supports are finely divided solids (125-250 μm or 60-120 mesh) of two general types: diatomaceous earth and fluorocarbon polymers. Two types of diatomaceous earth (or diatomites) are used: fire-brick derived materials that are usually used for non-polar samples, and filter-aid derived materials that are used for polar samples.

Table 2.6   COLUMN TYPES, DIMENSIONS, AND SAMPLE SIZES

Column Type	Internal Diameter	Stationary Phase Percent/Thickness	Normal Sample Size
Packed	6 mm	1-10%	1-100 μL
Packed	3 mm	1-10%	0.5-20 μL
Packed	1 mm	1-10%	0.01-0.1 μL
Capillary	0.25 mm	0.25 μm	$\leqslant$0.05 μL
Capillary	~ 0.35 mm	0.10 μm	$\leqslant$0.1 μL
		0.25 μm	$\leqslant$0.1 μL
		1.0 μm	$\leqslant$0.5 μL

The diatomites must be cleaned and deactivated before they can be used as supports. Since they are forms of silica, they have surfaces containing hydroxyl groups and, like glass, they are deactivated in the same manner. They are cleaned before deactivation by being washed with dilute hydrochloric acid followed by water and are then deactivated with trimethylchlorosilane or hexamethyldisilizane. Commercial acid washes and deactivated supports are available (Table 2.7). If the supports have not been completely deactivated, adsorption as evidenced by tailing of peaks, may occur. This problem can be alleviated by more careful deactivation or by a technique known as **priming.** In this technique, several samples are injected into the system (remembering that a GC column is reused many times) so that all adsorption sites are coated with the most polar component of the sample. This coating will be shown by the obtaining of uniform, symmetrical peaks. At this time, the real sample is injected and the desired measurements are made. Calibration is critical for quantitative measurements under these conditions.

The fluorocarbon or Teflon-type supports are used, frequently in Teflon tubing, when corrosive or highly polar compounds such as water are involved. In these cases, a fluorocarbon liquid phase may also be used.

Gas-solid chromatography, where there is no stationary liquid, is used to separate gases, such as air. In these cases such materials as molecular sieves, porous polymers, silica gel, activated carbon, and graphitized carbon serve as the stationary phase.

The mesh size of the support is less important although, as noted above, it is suggested that a narrow range of particle sizes be used to produce a uniform packing, which gives a minimum pressure drop. Possible sizes are 100-120 mesh for 2 mm columns, 60-80 or 80-100 for 3 mm columns, and 40-60 mesh for larger columns.

The Stationary Phase. The actual liquid phases have been considered previously and are summarized in Table 2.4. Amounts of phase and sample sizes are summarized in Table 2.6. Little needs to be added except perhaps for a discussion of the relative amount of phase present, or the degree of **loading.**

Obviously, enough phase must be present to completely cover the surface of the support and, thus, to dissolve the sample as it is being chromatographed. Beyond this point, the amount of stationary phase (thickness) will have an effect on the retention times as shown in equation 1.4. Increased amounts of phase will increase retention volumes and times. For qualitative work, a coverage of 2-5% is common with packed columns and a coating of 0.1-1.0 μm is common for capillary columns, with the qualification that, at low percentages, the support or walls must be carefully deactivated. For preparative work, enough phase must be present to dissolve relatively large amounts of sample, and loadings of up to 20% are common.

Table 2.7   SUPPORTS USED IN PACKED GC COLUMNS

Type	Name	Manufacturer	Column Use
Firebrick-derived[a]	Chromosorb P Gas Chrom R	Johns-Manville Alltech Associates	Non-polar compounds
Diatom-derived[a]	Chromosorb W Gas Chrom Q Supelcoport Anakron ABS	Johns-Manville Alltech Associates Supelco Analabs	Polar compounds
Molecular sieves[b]	Carbosieve Type 5A sieve	Supelco Linde Company	Gas analysis
Porous polymers[b]	Poropak Chromosorb 101–104	Waters Associates Johns-Manville	Highly polar compounds

[a] These supports are available as (1) non-acid washed, (2) acid washed, (3) acid washed and chlorosilyl treated, and (4) treated with trimethylsilane.

[b] Normally used without any added stationary phase.

Preparation and Packing of Columns. All column packings (support coated with phase) and columns of every size, shape, and tubing are commercially available. On the rare occasions when it is done, the preparation of a packed column involves two parts: the coating of the support with phase, and the placing of the packing into the column tubing.

The coating is carried out by dissolving the phase in an appropriate solvent, combining the solution with the prepared support and evaporation of the solvent so that the phase is deposited uniformly on the support. For example, 0.2 g of some non-polar phase such as a methyl silicone oil may be dissolved in hexane (any amount, but the hexane must be the purest available) and mixed with 10 g of support. The mixture can then be evaporated in a rotary evaporator with very slow rotation so that the particles will not be fractured. This will give a 2% loading. More specific directions are available for other situations in books in the Bibliography.

The packing is placed in the column in the following manner:

> One end of the column is plugged with silanized glass wool and the packing is added with gentle forcing (often vibration, but, better, vacuum pulling or pressure pushing) until the tubing is full, and then another plug of glass wool is inserted. Generally it is best to shape most metal and all glass columns before packing and a very long metal column after packing.

Capillary columns, that are usually purchased, pose some special problems. Since the tubing wall will be serving as the support for the liquid phase, it is quite crucial that it be clean. This is done by passing (vacuum pulling) a dilute solution of HCl through the column, followed by washing with water and drying with an inert gas. The walls of glass or fused silica are then sometimes coated with a preliminary layer of some polar material. Carbowax (a polyethylene glycol) is the current choice, although most metal and glass columns are also silanized with trimethylchlorosilane. These materials will assist in forming a uniform coating and decreasing adsorption. The actual coating with phase is accomplished by forcing (with inert gas pressure) or pulling (with vacuum) a solution of phase in some volatile solvent through the capillary tube. The concentration of the solution will determine the thickness of the layer. If the surface is not properly prepared, the phase will ball up, with the result that adsorption will markedly increase and efficiency will decrease.

Sometimes the wall of a larger capillary column is coated with a layer of finely divided support such as is used in packed columns. The phase is then placed on this support, resulting in a larger surface area and a greater amount of phase in the column. Such a system combines the advantages of packed and simple capillary columns.

Conditioning. After any column is prepared, it must be **conditioned.** This operation is, basically, the removal of any residual solvent and lower molecular weight stationary phase. The latter materials and their formation by slow decomposition of the stationary phase are the major cause of **column bleed.** In any case, the column is slowly heated in a GC oven (with carrier gas flowing) at a nominal rate to the upper limit of the stationary phase without the column being attached to the detector. A packed column should be maintained at that temperature for at least 24 hr. Certain stationary phases and very long packed columns take longer to condition. If excessive column bleed is noted at some specific temperature during use, as evidenced by an increasing and noisy baseline, the column should be used only at temperatures 25°C or so below that point. After conditioning, the column should be cooled to room temperature before it is removed from the oven. It is advisable to keep inert gas in the column at all times.

In addition to this conditioning of a newly prepared column, any column should be conditioned for a short period of time after it is placed in the instrument. This is done by heating the column, with the gas flowing at about 25°C higher than the planned maximum use temperature, for about 15 minutes. This will restabilize a column, especially a packed column.

Cleaning and Rejuvenation of Used Columns. After considerable use or misuse, columns can sometimes produce excessive amounts of tailing with an accompanying loss of efficiency. This situation can be improved by backflushing, steaming (injecting a small amount of water), and resilylation. A packed column often collects involatile materials on the packing, thus the first few centimeters of packing can be replaced. A capillary column often loses stationary phase at the end of the column; this can be alleviated by carefully cutting off the last 10 cm of the capillary. Some microprocessor-controlled GC's can be programmed to backflush after each injection, thus increasing column lifetime.

As mentioned above, when not in use, columns should be filled with an inert gas and closed at both ends. With proper care, a column can last a year or more.

Standardization and Characterization of Column Phases. A number of methods are available to measure the properties of phases against one another and to characterize, systematically, the degree to which phases retard certain types of samples. Some of these techniques are the so-called McReynold's Constants for various phases and the Kovats Indices. These concepts are summarized in books in the Bibliography. The use of the Grob test mixture for the characterization of capillary columns is described in recent books and is referenced.[2]

Mixed Phases. Thus far, we have considered only phases consisting of one type of compound or polymer. Mixed phases are also possible and, in fact, have some advantages. The most interesting aspect of such mixed phases is in the way that they can be used to optimize a given separation using a Window Diagram Procedure as worked out by Laub and Purnell.[3] In this technique, the sample is chromatographed on two different phases that bracket its polarity. The chromatograms are then analyzed using a simple computer program that will allow the prediction of certain mixtures of the two phases that will yield an optimum separation in the shortest time. The two phases can then be mixed and used in the same column, or placed in two columns in series.

Detectors

The detector is a device placed at the end of a GC column that analyzes the emerging gas stream and provides data to a data recorder which shows, graphically, the results of the chromatogram. Several of these were mentioned above and some description of the thermal conductivity detector (TCD) and flame ionization detector (FID) was presented. In this section, we will describe the mode of operation of these detectors and the electron capture detector (ECD), and discuss their maintenance.

Thermal Conductivity Detector. A TCD is based upon the fact that heat is removed from a hot body at a rate that depends on the composition of the gas surrounding that hot body. Thus, any given gas has a thermal conductivity. This conductivity is a function of the rate of motion of the gas molecules, which, at any given temperature, is a function of the molecular weight. Gases with the highest conductivity have the lowest molecular weight. When a substance of higher molecular weight is present (for example, any sample in a GC stream of helium), the conductivity of the gas is lowered, and the hot body is less effectively cooled.

The detector consists of a double arrangement of metal wires (platinum or some alloy) that can be used to 400°C. One wire is placed in the column effluent and one is placed in a reference gas flow (the carrier gas before it enters the column) at the same temperature as the column effluent. Currents are passed through the wires, thus heating them, and one is balanced against the other by a Wheatstone bridge. The sample in the column effluent reduces the conductivity of the gas stream allowing one wire to heat up. This signal appears as a difference in current from the reference, which is transmitted to a recorder or data system. The detector works best at a constant temperature and with good thermal insulation from the column compartment, especially when temperature programming is being used.

A major problem with the TCD is that the wires must be protected from oxygen while they are hot. Thus, they should never be heated unless the carrier gas is turned on. Many modern instruments have an interlock that prevents this problem. TCD's are usually cleaned by disconnecting them from the system and soaking them in a series of hot solvents such as decalin, methanol, water, and acetone. After drying, the detector is heated for 24 hr in the carrier gas flow of the chromatograph before use.

A major advantage of the TCD is that it is non-destructive, that is, the sample is not destroyed during detection. Thus, the TCD is ideal for preparative work or where the sample will be further characterized by mass spectrometry or infrared spectrophotometry (see below). The TCD is a concentration detector in that it measures the total number of moles passing through. The readings should be independent of flow rate.

Flame Ionization Detector. The basic idea in FID detection is that organic chemicals, when placed in a flame, break down to form simple, generally one-carbon, positively charged fragments. These fragments increase the conductivity of the flame area, and this conductivity increase is easily measured and recorded. Thus, the effluent gases from the column are passed into a hydrogen flame, which is burning in air, and two charged electrodes are placed in or around the flame. The sample, carried by the carrier gas, passes into the flame and is broken down to give ions. The ions increase the conductivity and thereby the current flowing between the electrodes. The current is amplified and recorded.

An FID measures the number of carbon atoms rather than the number of moles as measured by a TCD. The FID is essentially universal for almost all organic compounds (highly fluoro compounds and carbon disulfide are not detected) and highly sensitive. The response from an FID is quite linear (see Table 2.8) with respect to sample size, so that quantitative measurements are easy to make and accurate.

Most of the problems with an FID involve the hydrogen flame. The flow rate of hydrogen gas and the air to burn it must be correctly adjusted with respect to one another and to the gas flow from the chromatograph. If these are not in balance, the flame will be hard to ignite or to keep going. The air and hydrogen must be clean and dry, for example, water in the air supply will give a noisy and erratic baseline.

The design and cleanliness of an FID are critical. It can sometimes be cleaned by the injection of water followed by one of the Freons. More often it will have to be disassembled and cleaned in an ultrasonic bath with an appropriate solvent, usually again a Freon. Since the gas flow is being analyzed in a very hot flame, its actual temperature is of little importance; thus, an FID is insensitive to changing column temperature.

Electron Capture Detector. An ECD consists of a radioactive source, generally Ni^{63}, placed between two charged electrodes. As the carrier gas, nitrogen or argon plus methane, flows into the detector, it is ionized by the radioactive source to produce electrons, that then flow across the electric field to generate a current. When a sample is present in the gas flow it *captures* some of these electrons, thus reducing the current. The current reduction is then amplified and recorded. Since some organic functional groups are more efficient electron catchers than others, an ECD is more sensitive to certain types of compounds. Thus, halogen compounds, nitriles, nitrates, and highly conjugated systems are easily detected, whereas simple saturated hydrocarbons, ketones, and alcohols are not.

Where the TCD and FID are generally sensitive to all organic compounds, the ECD is much more discriminating. This can be useful for the trace analysis of certain compounds in dilute solution. For example, very small amounts of the chlorinated pesticides can be detected in the presence of large amounts of other compounds in environmental samples.

The ultra-high sensitivity of an ECD means that it is easily contaminated and that it will be difficult to clean. Since most ECD's cannot be opened for cleaning, they must be returned to the manufacturer, unless they can be cleaned by successive injections (of acetone, for example) or by a hydrogen purge at the maximum temperature of the ECD.

Others. A wide variety of GC detectors are available, many of which are specific for certain elements. Some of these are listed, along with those discussed above, in Table 2.8.

Dual Detection. Dual detectors can be used in series if the first one is non-destructive (such as the TCD). If both detectors are destructive, the gas stream must be split. Such systems are especially useful when one detector is element specific and one is universal. Thus, one can measure the amount of a given element in a specific compound and find out, at the same time, how much compound is present. The technique is also useful in detecting trace impurities.

Signal Handling

The amplified detector signal is recorded on a strip chart recorder to give a record such as that shown in numerous figures in this chapter. The major requirement for such a recorder is a fast response time (<0.1 sec). The signal can be decreased if necessary before being recorded by an **attenuator** in the GC and, after being recorded, can be used to obtain qualitative and quantitative data.

The Attenuator. The attentuator simply lowers the signal from the GC detector in a fixed manner (usually in steps that halve it) before it is recorded. It is most useful when a sample contains large amounts of some components and small or trace amounts of others. The recorder is usually adjusted so that the small peaks will appear at full signal. When large peaks emerge, they normally exceed the graph paper on the recorder. Thus, the signal is attenuated or reduced when a large peak elutes, and the amount of attenuation and the precise point at which the signal was attenuated must be carefully noted. When the instrument is controlled by a microprocessor, the attenuation usually can be done automatically. Also, the use of microprocessor-based data systems will permit automatic attenuation; see Figure 2.4. The result is a complete curve corrected for all attenuations.

Qualitative Data. GC data generally consist of retention times for the various components of any given mixture. The retention times are meas-

Table 2.8   LINEAR RANGES, DETECTABILITY, AND
             SENSITIVITY OF MAJOR DETECTORS

Detector	Linear Range, ng	Linearity
Thermal conductivity	90–700,000	$10^4 - 10^5$
Flame ionization	0.007–500,000	10^7
Electron capture		
DC	0.0007–0.07	10^2
Pulsed	0.0007–7	10^4
Photo ionization	0.001–100,000	10^7
Flame photometric		
S compounds	0.007–0.7	$10^2 - 10^3$
P compounds	0.007 (0.0001)–7(70)	$10^3 - 10^5$
Thermionic –		
P compounds	0.0007–0.7	10^4
N compounds	0.005–50	10^5
Hall electrolytic		
N	0.01–10	10^4
Sulfur	0.01–10	10^4
Halogen	0.005–500	10^6

ured from the point of injection to the peak maximum point and are quite characteristic of specific compounds *under specific conditions* (column, temperature, carrier gas, flow rate, etc.). Such data can also be provided on a digital print out, as shown in Figure 2.4, if the GC is so equipped.

The presence of a certain component in a given mixture can be confirmed by the spiking technique when a pure compound is available. The pure compound is added to the sample that presumably contains it, and the sample is chromatographed. If the appropriate peak is enhanced, in a symmetrical fashion, in two different stationary phase systems of different polarity, the component is probably present. Figures 2.12a and 2.12b show the technique.

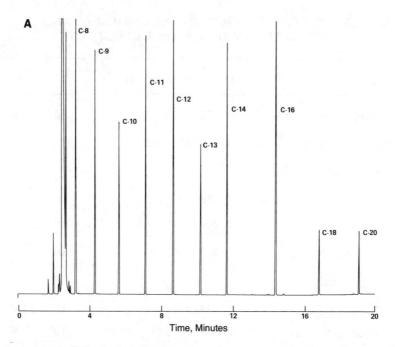

Figure 2.12 (a) A fused silica capillary column chromatogram of a mixture of C_5 to C_{20} hydrocarbons run as in Figure 2.6b and with the peaks labeled for identification purposes.

Quantitative Data.   An important aspect of GC and the HPLC discussed in Chapter 6 is that the data from a detector and as produced on a recorder can be quantified and, under the appropriate conditions, constitutes an accurate quantitative analysis. The actual measurement taken from the chart is the area under a given peak. If the peak is symmetrical or Gaussian, the height of the peak can be used for an area measurement. The area under the peak *or,* in some case, the height of the peak is proportional to the amount of compound present. Under ideal conditions and when microprocessors are involved, accuracies of less than 1% can be attained. Data from a recorder are in the 5-10% range of accuracies.

Measurement of Areas.   Integration of areas can be accomplished in several ways. The simplest is to cut out and weigh the paper peaks on an analytical balance. This can best be done on a copy recorder trace of the peaks. The weights can be converted to areas or simply used as such. Secondly, the areas can be measured with a **planimeter,** which is a mechanical device for measuring the area of any surface. Thirdly, the measurements may be made instrumentally. Some recorders have a Disc Integrator, which will give a measure of the area. The final method is electronic. Separate

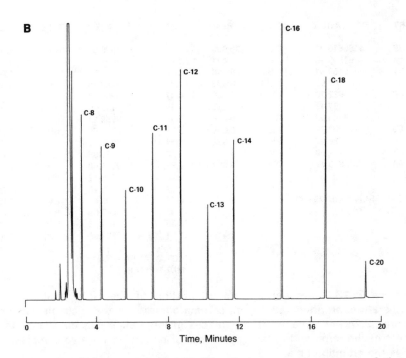

Figure 2.12 **(b)** A fused silica capillary column chromatogram of the sample in Figure 2.12a with C_{18} fraction spiked with additional material. Note the absence of any peak shouldering, supporting identification of peak as the C_{18} hydrocarbon.

electronic integrators are available, and the most modern ones utilize microprocessors. These include data integrators and complete computer systems. Examples of the output of such a system is shown in Figures 2.4 and 2.13.

Calibration. Since the areas under the peaks are only proportional to the sample amounts as sensed by the detector, it is necessary to establish a calibration before actual analytical data can be obtained. It is also necessary to establish that the calibration is accurate over the total range of possible results for the detector. See Table 2.8 for the ranges and limits of detector linearity.

The most accurate, as well as the most time consuming, way to make such a calibration is to make up a series of mixtures of known amounts of the sample components. These mixtures are chromatographed, and the curve areas are plotted against the known amounts to give calibration curves. Such curves allow one to measure the absolute amount of each mixture component as well as relative amounts. A less accurate measurement can be obtained through a spiking technique. A known amount of the compound(s) in question can be added to the mixture and chromato-

RT	AREA	TYPE	AREA %	RT	AREA	TYPE	AREA %
0.32	4186810.00	8BV	95.988	0.48	5664020.00	8VB	96.966
0.58	61999.40	BB	1.421	1.81	62004.40	BV	1.061
0.77	6459.81	BV	0.148	2.14	396.93	VV	0.007
0.86	69.02	VB	0.002	2.83	6460.85	BV	0.111
1.08	65682.70	BV	1.506	4.07	66208.60	BV	1.133
1.20	757.11	VV	0.017	4.46	709.48	VB	0.012
1.57	3769.08	BV	0.086	5.33	3758.11	BB	0.064
1.85	407.29	VB	0.009	5.83	380.65	BB	0.007
2.35	22788.60	BB	0.522	6.56	22886.80	BB	0.392
5.62	10127.50	BV	0.232	8.20	76.09	PP	0.001
14.03	2918.42	BB	0.067	8.89	10121.90	BV	0.173
				9.22	395.12	VV	0.007
				9.84	76.45	BB	0.001
TOTAL AREA = 4361790.00				10.38	96.43	BV	0.002
MULTIPLIER = 1				11.02	3207.00	BV	0.055
				11.36	161.05	VV	0.003
				11.51	267.90	VB	0.005

TOTAL AREA = 5841230.00
MULTIPLIER = 1

Figure 2.13 Computer system integration of the isothermal (left) and temperature programmed (right) chromatograms of the alcohols shown in Figures 2.2a and 2.6a. The differences in integration can be related to the effect of the large and differing amounts of solvent in the two samples. The RT is the retention time.

graphed. The area enhancement will then be due to the known amount of substance, *if the detector response is linear.* This method can also be considered as using an **internal standard.** The **external standard,** which the system is calibrated by other compounds, is the least accurate method when standards are used.

For simple cases where one wishes to measure the percent composition of some mixture, **correction factors** can be established. The sample is chromatographed and, as a first approximation, the areas of the peaks are assumed to be a correct measure of the relative amounts of the components. A synthetic mixture can then be made from known samples in the proportions and chromatographed. The responses can be redetermined and used to correct the first approximation.

Problems. The main problems in quantitative GC involve **sample discrimination** during injection and component interactions. If the sample is not completely volatilized in the injector or if some mixture components adsorb (more polar ones) or are deposited (higher boilers) in the injector, sample discrimination will result. It is also a fact that the calibration curve or correction factor for a compound will change when it is mixed with different compounds and in different amounts. Thus, it is important that all calibration and sample chromatograms be carried out in precisely the same manner.

2.5 SPECIAL TECHNIQUES

Sample Trapping

It is frequently desirable to trap one or several components of a mixture and use the pure compounds for further characterization. The simplest way to do this is to use a TCD (non-destructive) and trap the sample in a cooled piece of glass capillary tubing as it emerges from the detector. If an FID or an ECD must be used, a stream splitter will have to be incorporated so that only a portion of the gas stream goes to the detector. The capillary tube can be cooled with air, with ice, or with Dry Ice as appropriate, and should be connected just as the sample emerges and disconnected as soon as it has passed through. The procedure can be repeated to obtain larger amounts.

Preparative GC

Analytical GC can be scaled up to a point where large amounts can be separated. However, the process is not simple. Two possible ways are available for increasing the capacity of GC equipment: the use of a single large-bore column and the utilization of multiple separations with the regular or slightly increased column diameter. A non-destructive detector or a destructive detector with the splitter must be used.

Enlarged Diameter Column. This is a logical first step in increasing capacity, since a larger column will hold more liquid phase and be able to separate larger amounts. Columns of about 25 mm diameter will permit separations of an amount approaching 1 g. A large amount of carrier gas will be needed to maintain a satisfactory flow rate. Large columns are, of course, expensive and not easy to pack in a uniform manner.

Automatic Repeating Apparatus. The capacity of a given system can be increased by separating a large number of small samples and collecting the fractions automatically. Some equipment has been available (Varian Autoprep) that injects at timed intervals, and the detector response is used to trigger a rotating collection device to obtain products.

The ultimate in preparative GC has been worked out by Guiochon and co-workers[4] for the separation of the stereoisomers of *(meso* and *dl)* of 2,4-diphenylpentane. A packed column, 40 cm in diameter and 150 cm long, was used for the separation of successive 600 g samples. The samples were injected every 100 sec into the carrier gas of recycled hydrogen flowing at $60m^3$/hr. A total of 500 kg of sample was separated each day.

Multi-Dimensional GC

Each of the many chromatographic techniques, GC, LC, TLC, HPLC, and all the others, has its own special separation capabilities and specificities. These can be used in conjunction with one another to solve very complex separation and analysis problems. In GC, such methods consist of two columns containing different stationary phases and placed one after the other (in series). Presumably, those samples that cannot be resolved completely in a column of one type can be separated in the second column. It is crucial that the columns be connected with short tubing of the same diameter as the column and that no unswept volume be present to give excessive peak broadening or tailing.

Recycle GC

With an appropriate set of valves and efficient control mechanisms, it is possible to recycle the eluents of a single chromatogram back through the same column or to different columns pretty much at will. This **column switching** can be done with the columns in one oven or more than one. It is also possible to cut out certain fractions and recycle only these, or to chromatograph them in a separate column.

Pyrolysis GC

High melting, high boiling compounds and polymers (including biopolymers) cannot be examined by regular GC. However, if these substances are broken down or decomposed to smaller fragments that are volatile, they can be identified on the basis of these fragments. This controlled decomposition of solids or liquids and the chromatograms produced are the basis of pyrolysis GC.

The material to be pyrolyzed is usually placed either in a platinum boat or on a platinum wire in a commercially available attachment on the injection port of the regular GC equipment, usually one with an FID. The sample is heated in the carrier gas flow to the decompositon temperature (400-900°C) in a short period of time, either by passing a current through the platinum or by some other rapid heating technique. The mobile phase carries the pyrolysate onto the column that gives the chromatogram in the normal way. The temperature of the pyrolysis causes different types and extents of decomposition; thus, the temperature choice is important and must be reproducible for good results.

It is possible to relate the fragments produced from polymers to the monomeric units that make up the polymer, making a determination of the structure possible. The monomeric compounds will give the same retention time when injected in the normal manner.

Gas Chromatography/Mass Spectometry

A mass spectrometer has the capability of analyzing very small samples and producing from them enormously useful data on the structures and identities of organic compounds. If the effluent from a GC is directed into

Figure 2.14 Picture of a gas chromatograph/mass spectrometer/data system. (Reproduced through the courtesy of Finnigan-MAT.)

a mass spectrometer, such structural information can be obtained for *each peak* in the chromatogram. Because of low flow rate and small sample size, this technique is easiest with a capillary GC column.

A detailed discussion of this technique is beyond our scope, but we can describe what an instrument such as the Finnigan-MAT 4500 system, as shown in Figure 2.14, can do. The sample is injected into the GC and undergoes chromatography so that the components are resolved. A mass spectrum is automatically measured at timed intervals or at the maximum or centers of the peaks, as they emerge from the column. The data are then stored in a computer. One can obtain, on request from the computer, the results of the chromatogram, with the integrals under all of the peaks. Secondly, one can get a mass spectrum of *each* of the components. Such a spectrum can serve as an absolute method of identification for known compounds and as a source of structural information and molecular weight for unknown compounds. Finally, many such instruments are equipped with computerized libraries of the mass spectra of thousands of known compounds. On request, the computer will search its library for any spectrum similar to the one obtained for a given peak and suggest a compound or compounds if the spectrum is in the library. Such instruments have revolutionized trace analysis, flavor analysis, and much of organic chemistry. However, they are expensive and require expert operators.

Gas Chromatography/Infrared Spectrophotometry

Instrumentation is available for using an infrared spectrophotometer in the same manner as the mass spectrometer. This technique used to be complicated by the fact that sample size must be fairly high for infrared measurements and that it takes several minutes to make a typical infrared spectrum scan. However, Fourier transform infrared spectrophotometers have made it possible to solve both problems. The FT/IR gives information on molecular structures, as does GC/MS, and, in addition, provides information on structural isomers not possible by GC/MS.

2.6 OPTIMIZATION OF A SEPARATION

General Factors

In Chapter 1, we considered an ideal or theoretical chromatogram and described how such parameters as the number of theoretical plates, the relative amounts of stationary and mobile phase, and the partition coefficients affected the retention volumes (LC) and retention times (GC) and the broadness of bands or peaks. Thus, some band broadening is *inherent* in the chromatographic process.

We then considered the Van Deemter equation, 1.8, that is a sort of catch-all equation describing deviations from a theoretically perfect chromatogram.

$$H = A + \frac{B}{\bar{v}} + C\bar{v} \qquad (2.1)$$

In equation 2.1, H is the height equivalent to a theoretical plate, and smaller numbers mean a more efficient column. Three of the terms that make up the equation are A, a factor for eddy diffusion which is the undesirable flow of mobile phase in the column; B, which is spreading of sample due to diffusion (separate from the natural spreading due to the chromatographic process); and C an equilibrium term describing just how complete an equilibrium exists in the column. In essence, these three terms are a measure of the perfectness of the column packing or coating. In general, these factors are minimized by using a finely divided support with a narrow particle size range and a correct loading of liquid phase (the completeness of the coating in a capillary column, as well as its thickness). However, the average velocity, $\bar{v}$, is also included in the Van Deemter equation and has a direct effect on the diffusion and equilibrium terms. Unfortunately, the effects are opposite with a higher velocity reducing the diffusion term (to give a more efficient system) and increasing the equilibrium term (leading to a less efficient system). Thus, for any chromatography there is a optimum velocity that is generally determined by trial and error, as discussed above under gas flow rate.

Measurement of Separations

Thus far, we have defined GC separations only in terms of retention times, that is, the time that a sample is retained on a column. Such times are dependent upon a number of factors. Some of these, such as the nature of the mobile and stationary phases, are, in a sense, given. However, some of these factors, such as the length of the system and the volume of the mobile phase in the column, can be compensated for. This is done in the **separation faction,** k', as defined in equation 2.2 where t_R is the observed

$$ k' = \frac{t_R - t_M}{t_M} \tag{2.2} $$

retention time and t_M is the retention time of some material that is not retained by the column. In GC using a TCD, this is often simply air. The term k' is, therefore, a property of any sample component independent of the column length but, of course, dependent on the basic nature of the column. The best chromatographic results are obtained when the k' of a given components is about 5.

Speeding Up of a Chromatogram

The following ways of speeding up a separation are listed in a random order. Many, if not most, are used together, so that a number of small changes can greatly increase the speed of a chromatogram.

1. Higher column temperature. An increase in temperature of 30°C doubles the rate of throughput (but will decrease resolution).

2. Decrease in amount of stationary phase. The decrease in amount is not linear with throughput, but is related.

3. Increased gas flow. The gas flow may be less than the optimum amount.

4. Programmed temperature techniques. These are often used not only for speeding up a chromatogram, but also for getting a better separation by peak compression.

5. Dual column techniques. These permit a lower temperature and flow rate to be used to obtain the same separation.

6. Capillary columns. These will often give a separation in seconds for which a packed column may requires minutes.

REFERENCES

[1]C.L. Stong, *American Scientist,* June **(1966).**

[2]K. Grob, Jr., G. Grob and K. Grob, *J. of Chromatog.* *156* **(1978)** 1.

[3]R.J. Laub and J. H. Purnell, *Anal. Chem.* *48* **(1976)** 799.

[4]B. Roz, R. Bonmati, G. Hagenbach, P. Valentin and G. Guiochon, *J. of Chromatog. Sci.* *14* **(1976)** 367.

Chapter 3

CHOICE OF A LIQUID CHROMATOGRAPHIC SYSTEM

3.1 INTRODUCTION

In Chapter 1 we introduced the major types of chromatography and a number of the techniques of chromatography: TLC, GC, HPLC, etc., along with a brief description of each. We also presented a discussion of the kinds of information that one may seek from chromatography: qualitative, quantitative, or preparative results.1 GC was discussed in detail in Chapter 2. In this chapter we will consider the various methods and philosophies that may be involved in choosing a liquid chromatographic method (LC) for the separation of a mixture.

The actual choice of an ideal system for a given purpose is not easy. Should one use liquid-solid chromatography (LSC) or a liquid-liquid chromatography (LLC) system? Which adsorbent or support should be used? What solvent or mixture of solvents? When should GC be used? Partial answers to these questions lie in the literature, in experience, in the equipment and expertise available, and in an understanding of some of the basic phenomena involved. Complete answers do not exist. The remainder of this chapter will be devoted to a discussion of the concept of polarity, the choice between LSC and LLC, and the basic phenomena underlying these concepts. In addition, the manipulation of the LC systems, methods for scouting a possible method, and a comparison between some of the characteristics of the various LC techniques will be presented.

3.2 POLARITY

The phenomena that underlie most of the chromatographic techniques are, as described in Chapter 1, solubility, adsorption, and volatility (Figure 1.1 and associated discussion). These phenomena result from the tendency, or lack thereof, of molecules to associate with one another and can be best understood in terms of the **polarities** of the molecules involved. In essence, polar molecules tend to associate with one another (dissolve or adsorb), and

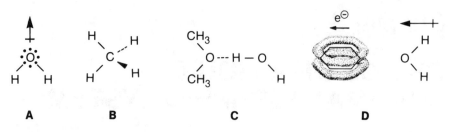

Figure 3.1 Molecules that show different types of polarities: (a) water-showing a dipole, (b) methane-representing a symmetrical non-polar system, (c) methyl ether-showing a hydrogen bond to water, (d) benzene and water-showing induced polarization.

non-polar molecules tend to associate with their own kind, or *"like likes like."*

A polar molecule is one in which the electron center or negative center of a molecule is some finite distance from the positive center (due to the differences between the electronegativities of atoms and the bond angles). The classic example of a polar molecule is water (Figure 3.1a), in which the electrons are displaced toward the electronegative oxygen, resulting in a **dipole,** as shown. Molecules having such dipoles will tend to attract one another like a group of small magnets. Methane, however (Figure 3.1b), is a completely symmetrical molecule in which the positive and negative centers are superimposable and which has no dipole and, thus, is non-polar.

In chromatography the term polarity is broadened to include hydrogen bonding (leading to a high degree of "polarity") and induced polarity (leading to polarity where there should be none). For example, water has a dipole and is hydrogen bonded, making it much more polar than dimethyl ether, Figure 3.1c, that has a similar dipole but is not hydrogen bonded. While it is generally understood that hydrogen bonding accounts for the strong attraction between molecules of the same compound, it is not always clear that the abilities of molecules to hydrogen bond account for the attraction between molecules of different substances. For example, in Figure 3.1c, a hydrogen bond is shown between water and the oxygen of dimethyl ether, leading to an appreciable solubility between the two. Thus, molecules that can hydrogen bond tend to attract one another and are said to be polar.

The phenomenon of induced polarity is less straightforward and is illustrated in Figure 3.1d. Benzene is a completely symmetrical molecule that has no permanent dipole and that should therefore be completely non-polar. This is not the observed situation, since benzene in mixtures behaves like a somewhat polar molecule (it is slightly soluble in water). This polarity arises by an induction process. As the water molecule approaches the benzene molecule, the relatively loose *pi –* electron cloud is pushed or pulled from its equilibrium condition to give rise to a temporary or induced dipole.

The relative polarities of liquids are described by their **dielectric constants** that can be measured directly.

Although the discussion above is given in terms of liquids, it should also be appreciated that the concepts apply equally well to solid surfaces. Thus, adsorbents in chromatographic systems may also be polar or non-polar. The two main adsorbents used are silica gel $[(SiO_2)_x]$ and alumina $[(Al_2O_3)_x]$ and, by virtue of the polar oxygen atoms as well as surface hydroxy groups, they are quite polar materials. Thus, both adsorbents will attract or adsorb polar molecules more tightly than non-polar molecules. Charcoal, which is used as an adsorbent for many applications (gas masks and various other poison and odor removing systems) and which can also be used as a chromatographic adsorbent, is an example of a non-polar adsorbent. Thus, since "like likes like," non-polar molecules are attracted more readily to charcoal than are polar molecules.

The relative polarities of some common solvents are given in Table 3.1. This is the so-called eluotropic series, which describes polarity as it is

Table 3.1   ELUOTROPIC SERIES OF SOLVENTS

Hydrocarbons (<u>petroleum ether</u>, hexane, heptane, etc.)

Cyclohexane

Carbon tetrachloride

Benzene[a]

<u>Toluene</u>

Methylene chloride

Tetrahydrofuran

Chloroform

<u>Ethyl ether</u>

Ethyl acetate

Acetone

n-Propanol

Ethanol

Acetonitrile

<u>Methanol</u>

Water

[a] *Benzene is a known carcinogen; thus, extreme care must be taken when it is used. Toluene is slightly more polar than benzene and is not a carcinogen.*

manifested in chromatography on a polar adsorbent such as silica or alumina. In essence, the solvents are listed according to their ability to compete with a polar surface for solute molecules. Those solvents at the bottom of the list are quite polar, and would tend to compete favorably with the surface, and readily move the solute through the system. The eluotropic series is not exactly the same with respect to various adsorbents, but is enough so to be invaluable.

The term polarity also has a bearing on the solubility effects that underlie liquid-liquid chromatography and some of the liquid phases that are used in GC. In this case "like likes like" becomes "like dissolves like;" polar solvents tend to dissolve polar solutes and vice versa.

In general, the polarity of organic compounds increases with the number of functional groups and decreases with an increasing number of carbon atoms. An interesting exception to these generalities is the polarity of the various perfluorohydrocarbons (CF_4, C_2F_6, C_3F_8, etc.), which are less polar than the corresponding hydrocarbons. Polarity is also used to describe the properties of molecules containing actual positive and negative charges. Thus, salts of acids and bases and the amino acids (in their zwitterionic form) are considered to be very polar molecules.

3.3 LIQUID-SOLID vs LIQUID-LIQUID

The various techniques or modes of LC may involve either LSC or LLC depending on whether the stationary phase is a solid or a liquid. Paper chromatography is always LLC; TLC and column chromatography are usually LSC; and HPLC can be used in either mode.

The choice between LSC and LLC will be governed by three factors: the relative experimental difficulty of the two methods; the purpose and goal of the chromatography; and, primarily, by the type of compounds being separated.

LLC processes are dependent upon a competitive solubility in two liquids and are quite sensitive to small differences in molecular weight between solutes being separated. For this reason, the members of a homologous series are generally best separated by an LLC method, especially those members containing more than four or five carbon atoms.

LSC processes, on the other hand, are quite sensitive to steric or spatial differences between solutes. The extent to which the solute can be accommodated on an adsorbent surface will depend upon its configuration, and this, in turn, will determine its LSC relative to other solutes. LSC processes also depend upon the ability of a solute to hydrogen bond with the adsorbent surface. In summary, LSC methods are probably best for similar molecules with a slightly different shape, for molecules having different numbers of negative atoms such as oxygen or nitrogen, or for molecules with different functional groups.

In the past, LSC has been experimentally easier than LLC chromatography. However, the use of the bonded phase columns in HPLC has made the technique much easier. In LSC, to a first approximation, the adsorbent surface is constant, and the solvents are varied over the eluotropic series (Table 3.1) in search of a polarity that will cause a separation. In LLC chromatography, neither of the two phases can be maintained at constant polarity, and it is more difficult to carry out systematic experimentation. However, the use of HPLC with gradient possibilities and a bonded phase column allows a more controlled polarity change.

LSC can be used for the separation of larger samples than can LLC. Usually the adsorptive surface in the highly porous absorbent particles can take on a larger amount of solute than the more limited amount of stationary phase in a LLC system.

LSC, particularly as practiced in columns, can be used for the separation of samples containing solutes of widely differing polarities. In theory and in fact, one can place a sample on an adsorbent and, using solvent mixtures, expose the sample to the total range of solvent polarity as given in the eluotropic series. LLC, in contrast, is best applied to mixtures of fairly similar solutes. Bonded phase partition chromatography can involve gradient elution, but has solvent limitations not present in LSC. It is quite common to carry out preliminary large-scale fractionations of crude mixtures by an LSC method with a wide range of solvent polarities and to submit these crude fractions to LLC for final separation. As a final comment, LLC generally has a higher resolving power than LSC and can be, under the proper conditions, a better quantitative method.

It would seem logical to assume that any mixture could be separated by LSC if one could find a solvent or mixture of solvents having the correct polarity. Sad to say, this is not quite true. When solutes are polar and tightly adsorbed to a surface, it is necessary to use highly polar solvents to move them. Frequently, these highly polar solvents will overwhelm small differences between solutes and produce no separation. As a general rule, highly polar materials such as carbohydrates, amino acids and nucleotides are separated by LLC methods, and relatively less polar molecules such as mono- or difunctional organic compounds are separated by LSC. Keeping in mind that polarity depends upon the nature and number of functional groups, it is possible to assemble the diagram shown in Figure 3.2, which might help in the choice of a system.

In Figure 3.2, the polarity increases from left to right, primarily due to increased dipoles in the functional groups involved and their increasing ability to hydrogen bond, and from top to bottom depending upon the number of functional groups. Thus, the most polar substances (those most amenable to LLC) will be found in the lower right-hand corner of the diagram and the least polar (most amenable to LSC) will be found in the upper left-hand corner. A diagonal dashed line from lower left to upper

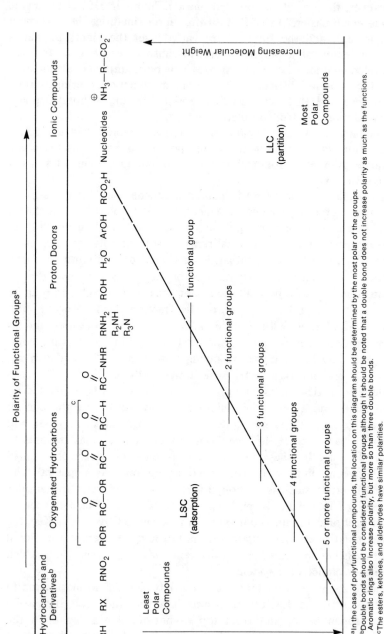

Figure 3.2 Liquid–solid (adsorption) vs liquid–liquid (partition) chromatography.

[a]In the case of polyfunctional compounds, the location on this diagram should be determined by the most polar of the groups.
[b]Double bonds should be considered functional groups although it should be noted that a double bond does not increase polarity as much as the functions. Aromatic rings also increase polarity, but more so than three double bonds.
[c]The esters, ketones, and aldehydes have similar polarities.

right divides the diagram into those compounds that should be separated by the two methods. Note that the line begins (at the bottom) to the right of the hydrocarbons and ends (at the top) to the left of the amino acids and nucleotides. This would indicate that the amino acids and nucleotides should always be separated by LLC and hydrocarbons be separated by LSC. While the former is true, the hydrocarbons constitute a special case. Saturated aliphatic hydrocarbons are so non-polar that they will not adhere to the typical polar adsorbents and no separation can be made. They are best separated by GC because of their volatility.

In actual fact, there is a broad range of substances for which either of the methods may be used, depending on other factors. Also, it is not entirely clear whether LSC or LLC or a mixture of both is taking place in a given experiment. The increasing use of reversed phase bonded partition chromatography is adding another variable to the capabilities of liquid chromatography.

3.4 LIQUID-SOLID CHROMATOGRAPHY PROCESSES

Nature

Two competing solute properties are involved in LSC: the adsorbing properties of the solutes that tend to hold them onto the adsorbent surface and the solubility properties of the solutes that tend to allow them to dissolve in and move with the mobile phase and be separated. The factors that determine both of these properties were summarized in Section 3.2.

The chromatography of acids and bases is somewhat different in that one often has ionic forces playing a role. Of the adsorbents discussed in detail in this book, one (silica gel) is relatively acidic and the other (alumina) is relatively basic, at least in their unmodified forms. If acids are chromatographed on alumina, they will be tightly bound to the surface by ionic forces and will be difficult to move and resolve. The same is true of bases on silica gel. Thus, bases should be chromatographed on alumina, and acids (and phenols) should be chromatographed on silica gel, at least in an initial effort.

There are actually two phenomena that may take place when a solute moves with the mobile phase in a chromatogram. The first of these is the one that has been discussed in detail above, that is, the tendency for a solute to dissolve in and move with the solvent. In this case, the ideal solvents should dissolve the solutes and should be just good enough, as solvents, to compete with the adsorptive power of the adsorbent. This situation probably prevails when non-protonic solvents such as hydrocarbons, ethers, and carbonyl compounds are being used as developing solvents.

The second phenomenon involved in the movement of solutes in a chromatogram is **displacement.**   The solvent molecules tend to compete with solute molecules for sites on the adsorbent surface and to move them by pushing them out of these sites.   The type of chromatography based on this notion is called **displacement analysis,** and the displacing molecules may be solvent or even another reagent dissolved in the solvent for the express purpose of displacing the original solute.   The introduction of protonic solvents such as alcohols or amines almost surely produces some displacement.

The line of demarcation between these two phenomena is quite blurred. Most separations can be carried out and understood if it is assumed that only elution chromatography is taking place and that the protonic solvents are merely very polar.

One of the major properties of an adsorbing surface is its **degree of activity.**   Normal, unactivated adsorbents have a surface completely coated with water.   That is to say, the adsorbent sites mentioned above are all occupied by very polar and hydrogen bonding water molecules.   Some of these must be removed by heat or **activation,** generally at 100 to 110°C, to produce an activated adsorbent before chromatography can be carried out.

Manipulation - LSC

LSC processes can be manipulated by changing the nature of the adsorbent surface or by changing the polarity of the solvents.   Solvent manipulation is much easier and is generally used.

Although solvent manipulation is easier, a major variation does lie in the type of adsorbent chosen.   As noted above, acidic materials should be separated on silica gel and basic materials should be separated on alumina. Neutral materials can be separated on either, although one will frequently produce better separations.   Silica gel is more often used in TLC whereas the classic column adsorbent has been alumina.   After this preliminary choice, the properties of the adsorbents can be varied further.   Alumina can be prepared in several activities and in forms that are more or less basic and even acidic.   This is frequently done in column work and will be discussed in detail in Chapter 5.   Silica gel can be similarly modified in many ways to enhance separations.   These have been highly developed in TLC and will be considered in detail in Chapter 4.

Solvents are manipulated primarily by varying them and mixing them to produce an appropriate polarity for a given separation, generally using the eluotropic series (Table 3.1) as a guide.   The use of the eluotropic series is not completely straightforward.   Certain solvents or mixtures seem to have some special properties for the separation of certain solutes.   For example, one might prepare a blend of toluene and ethyl ether, which has the same polarity as chloroform (Table 3.1), but find that chloroform produces a better separation. Such matters require experimentation, experience, and luck.

Three factors are important to remember when one mixes solvents to produce a blend. The first of these is that only solvents having fairly similar polarities should be blended. For example, from Table 3.1 it would be possible to produce a blend having a polarity similar to that of chloroform in an almost infinite number of ways. One could mix cyclohexane with acetone, or carbon tetrachloride with ethyl acetate, and so on. The best way would be to mix tetrahydrofuran or dichloromethane with ethyl ether, since these solvents have polarities not too different from one another. If the pure solvents have widely differing polarities, tiny differences in the mixture composition will produce large differences in polarity, and different types of chromatography (elution and displacement) may even take place.

The second factor to remember is that the polarity of a mixture is *not* a linear function of the composition but a logarithmic function. Thus, the first tiny amount of ethyl ether added to toluene changes the polarity appreciably, whereas the difference between a 50% and a 60% mixture is quite small. The polarities produced by blending solvents are uniquely shown in Figure 3.3. Note, for example, that a 10% solution of cyclohexane in acetone is almost half as polar as a 50% solution (see area in dashed circle in Figure 3.3). The data in Figure 3.3 were measured by Neher and von Arx[1] and are based upon the Rf values of 20 steroids on silica gel. The polarity increases from left to right in the figure and any vertical line (such as the dashed line) will connect mixtures of similar polarities.

Finally, it should be noted that one can use a gradient between two solvents in some methods. One begins the chromatogram with one pure solvent and gradually mixes in the second during the chromatography, thus producing a continually increasing polarity to give **gradient elution.** This is discussed in detail in Chapters 5 and 6.

3.5 LIQUID-LIQUID CHROMATOGRAPHY PROCESSES

Nature

In LLC, the two competing phenomena are the solubilities of the solute in two liquid phases that are not completely soluble in one another. From this lack of mutual solubility, it follows that one phase must be much more polar than the other. Either liquid may be pure or a mixture of various materials.

The general aspects of solubility have been considered in Section 3.2 on polarity and Section 3.4 on LSC above. As described in Chapter 1, the

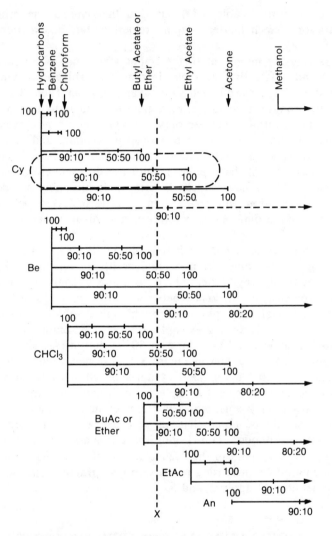

Figure 3.3 Solvent and binary solvent mixtures arranged in an equi-eluotropic series on the basis of average Rf values for 20 steroids. Any vertical line such as the dashed line x will connect solvent mixtures having an equal ability to move solutes in an LSC system. The dashed circle area shows the differences in polarity when solvents are mixed. Pure methanol is somewhere out of the figure to the right. Solvent abbreviations are: Cy=cyclohexane, Be=benzene, BuAc=butyl acetate, ETOAc=ethyl acetate, Ac=acetone. (Redrawn from Neher.[1])

extent to that a solute will dissolve in, or distribute itself between, two liquids can be measured in a separatory funnel. Such measurements produce **distribution coefficients.**   Materials having different coefficients are theoretically separable. Those solutes that are more soluble in the stationary liquid will move more slowly in a chromatographic system, and those that are more soluble in the mobile phase will move faster.

Although an LLC system is theoretically simpler than an LSC system, a major experimental problem is caused by the requirement that the stationary liquid phase must be held securely in place on some type of support while the mobile phase flows over it.   In the past, this has been accomplished primarily by carefully saturating the mobile phase with the stationary liquid before chromatography.   If the mobile phase is thus saturated, it will not dissolve and remove the stationary liquid. This is done by shaking the two phases together in a separatory funnel. More recently, the problem has been solved using liquid phases that are chemically bound to a support, the so-called bonded phases. These were developed primarily for HPLC and will be discussed in Chapter 6. However, they have been used in all types of LC and GC. Since it is easy to exceed the solubility of solute in the relatively small amounts of stationary phase present (at least relative to the mobile phase), LLC has a lower capacity than LSC.

Manipulation - LLC

An LLC system is manipulated by changing the nature of the two liquid phases, generally through the addition of other solvents or substances. In most cases, the more polar of the two liquid phases is water. The tendency of water to dissolve solutes can be changed by adding salts to produce a salting-out effect, buffers to dissolve or precipitate acids or bases, complexing agents to produce a specific type of solubilizing effect for a given functional group or class of compounds, or a miscible organic solvent. The less polar stationary phase liquid in an LLC system is usually methanol, but can be any organic compound such as acetonitrile, tetrahydrofuran, ethyl ether, or n-butanol, which have a limited solubility in the mobile phase. The ability of these organic stationary phases to dissolve solutes can be varied by adding more or less polar liquids or by adding organic acids and bases.

The problem, however, is rarely this simple, because the two liquids must be in contact and in equilibrium with one another. Thus, any material that is added to the mobile phase in order to influence its dissolving properties will tend to distribute itself between the two phases and change the properties of *both* phases. For example, suppose that a water-toluene mixture is the basic starting point for a system. In order to increase the ability of the toluene to dissolve polar solutes, ethanol is added. The ethanol will not remain entirely in the toluene, however, but will partially move into the water phase, thus changing the properties of both the toluene and the water phases. This situation is in sharp contrast to an LSC system where one can hold the adsorbent surface fairly constant and vary the polarity of the mobile phase.

3.6 OTHER FACTORS IN THE CHOICE OF AN LC SYSTEM

In the preceding sections, we have tried to explain the concepts that underlie the two major types of chromatography, so that one can logically choose a system to make a desired separation. However, such a choice is rarely made in a vacuum, and many other factors play a role. Some of these are the type of information or results desired (qualitative, quantitative, preparative), the equipment and money available, and, to a great extent, the type of experience and expertise available.

We have attempted to compare some of these many factors with one another for the various techniques. These generalizations are given in Table 3.2. Table 3.2 reflects our experience and opinions, which may vary widely from those of others.

3.7 SPECIFIC CHOICE OF EXPERIMENTAL CONDITIONS

The search for specific absorbents, liquids, or bonded phases for a specific technique and especially for the solvent mixture to be used is sometimes called **scouting.** Ideally, such scouting should involve the ability to investigate a number of systems either simultaneously or within a short time with a minimum expenditure of adsorbents and solvents. Clearly, the best technique to use in scouting an LSC or a LLC system for a non-volatile but soluble solute is TLC. Individual experiments require only a few minutes, and a number of TLC determinations can be carried out simultaneously in different chambers. The extrapolation of TLC results to column chromatography and to HPLC is theoretically simple, but practically less so. However, it can be done and will be the subject of discussion in Chapter 4.

Our approach, thus far, has been that one is dealing with a mixture whose general properties may be known, but whose actual chromatographic behavior has not been recorded. Actually, many thousands of chromatographic separations have been described in the chemical and biochemical literature, and many of them are summarized in the texts cited in the Bibliography. If these sources are available, it would be wise to check them before embarking on an experimental approach. If the mixture is of unknown compounds or if the method of choice is LSC, it is frequently about as easy to find a system by experimentation as it is to look one up. In

Table 3.2   TECHNICAL COMPARISON OF MAJOR CHROMATOGRAPHIC TECHNIQUES

	TLC	PC	CC	HPLC	GC
Chromatography type	Mostly LSC	LLC	Mostly LSC	LLC and LSC	Mostly gas liquid
Sample size	1 µg – 1 g	1 – 10 µg	1 – 25 g	10 µg – 10 g	1 – 50 µg
Minimum sample detectable	0.001 µg	0.001 µg	---	10^{-7} µg	10^{-9} µg
Approximate theoretical plates per meter	few	many	few	75,000	4000
Mobile phases	infinite (mixtures)	infinite (mixtures)	infinite (gradient)	infinite (gradient)	few
Stationary phases	~5	infinite	~5	~5 – 10	many
Applications:					
Qualitative	good	good	poor	good	good
Quantitative	poor (5-10%)	poor (5-10%)	bad	good (0.01%)	good (0.01%)
Preparative	fair	bad	good	good	poor
Sample complexity	2 – 10 comp.	2 – 10 comp.	2 – 20 comp.	2 – 100 comp.	2 – 1000 comp.

Table 3.2, continued

	TLC	PC	CC	HPLC	GC
Experimental complexity	simple	simple	moderate	complex	complex
Equipment cost	inexpensive ~ $100	inexpensive ~ $50	moderate $100–500	expensive $4000–25,000	expensive $4000–25,000
Learning time	1–3 days	1–3 days	1–2 months	2–3 months	2–3 months
Time for one separation	30 minutes	5–15 hours	3 hours–days	2–200 minutes	2–200 minutes
Sample purity required	crude	partially purified	crude	partially purified	partially purified
Type of detection	visual	visual	visual	instrumental	instrumental
Degree of automation possible	not possible	not possible	difficult	common	common
Reproducibility of results	fair	good	fair	good	excellent

contrast, if one is dealing with an LLC system, it is a good idea to check the literature carefully, since such systems are not generally easily devised experimentally. However, in HPLC the use of a bonded phase column and a gradient elution program utilizing methanol/water from 0-100% often will define a correct solvent mixture for a complete separation. The extensive literature on paper chromatography is especially useful for background on LLC systems.

Scouting an Adsorption Method

An experimental approach to this problem can be outlined as follows, based on the initial solvent mixtures described in Table 3.3.

Table 3.3   SOLVENT MIXTURES FOR TLC SCOUTING
RUNS ON SILICA G

Compound Class	Developer	Visualization
Long chain aliphatic alcohols	Pet ether/EtOEt/HOAc (90:10:1)	H_2SO_4 + heat
Long chain aliphatic ketones	Pet ether/EtOEt (9:1)	2,4-DNPH
Alkaloids	CH_2Cl_2/EtOEt	Dragendorff's reagent
Phenols	Hexane/ethyl acetate (4:1)	Diazonium salt
Polynuclear aromatics	Hexane/ethyl acetate (3:2)	U.V.
Food dyes	MEK/HOAc/MeOH (40:5:5)	Colored
Lipids	EtOEt	H_2SO_4 + heat
Steroids	Cyclohexane/ethyl acetate (9:1)	$SbCl_3$ in $CHCl_3$
Plasticizers	Isooctane/ethyl acetate	I_2

Prepare or purchase a series of thin layers containing the adsorbents of interest, generally silica gel or alumina. The directions for layer preparation are given in Chapter 5 and sources of commercial layers are given in the Suppliers List in the back of the book. The homemade layers should be activated with heat before being used; see Chapter 4. In case one is scouting for a column system or an HPLC system, some care should be taken to obtain a TLC adsorbent as similar as possible to the column packing to be used. Spot the mixture to be separated on four layers and develop them in the four solvents underlined in Table 3.1: hexane, toluene, ethyl ether, and methanol. Visualize the chromatograms if the solutes are colorless and compare the results. In this manner, one may find a pure solvent that will resolve the mixture, or at least learn something of its general properties. Frequently, it is possible to bracket the properties of the mixture, that is, to find one solvent that will move the spots too far and one that will not move them far enough. Mixtures of these two solvents can then be investigated as well as other pure solvents that lie between them in Table 3.1. In general, the best separations are obtained when the solute spots run less than halfway up the layer (see Chapter 4).

If streaking and poor separations are observed, there are some additives that can be put into the solvent to prevent them. When acidic or basic materials are being chromatographed on silica gel, a drop of acetic acid (for acids) or ammonium hydroxide or diethylamine (for bases) in the solvent will sometimes help. These reagents buffer the acidic or basic solutes to keep them in the un-ionized form so they will give more compact spots.

Some applications of adsorption chromatography in TLC are given in Table 3.4. These are taken from the book by Bobbitt on TLC,[2] but many others are cited in the literature. These examples should give some idea of the general conditions that will give satisfactory separations. The visualization reagents are those used in TLC. Additional suggested methods are given in parentheses.

Scouting a Partition Method

If a literature search fails to turn up a partition system that has been used to separate a specific mixture or a similar one, it may be necessary to devise one by experimentation. A number of partition systems are given in Table 3.5. Reversed phase LLC is defined in Chapter 6. An experimental system using unbound water as the stationary phase and various organic liquids as mobile phases can be defined as follows:

Table 3.4   SUGGESTED SYSTEMS FOR ADSORPTION CHROMATOGRAPHY

Compounds	Adsorbent	Developer	Visualization[a]
1. Long chain (over eight carbons) aliphatic ketones	Silica gel G	a. Benzene: EtOEt mixtures b. Toluene: EtOEt mixtures c. Pet ether: EtOEt mixtures	Phosphomolybdic acid (2,4–DNPH)
2. Aliphatic and aromatic aldehydes and ketones	Aluminum oxide	a. Toluene b. Toluene: EtOH (98:2) (95:5) (90:10) c. $CHCl_3$ d. EtOEt e. Pet ether: toluene (1:1)	2,4–DNPH
3. Aromatic aldehydes and ketones	Silica gel G	Hexane: EtOAc (4:1) (3:2)	2,4–DNPH (U.V. on phosphors)
4. Vanillin and other substituted benzaldehydes	Silica gel G	a. Pet ether: EtOAc (2:1) b. Hexane: EtOAc (5:2)	2,4–DNPH (U.V. on phosphors)

Table 3.4, continued

Compounds	Adsorbent	Developer	Visualization
		c. $CHCl_3$: EtOAc (98:2) d. Decalin: CH_2Cl_2: MeOH (5:4:1)	
5. 2,4-Dinitro-phenyllhydrazones of aldehydes	Silica gel G	a. Toluene: pet ether (3:1) for aliphatic b. Toluene: EtOAc (95:5) for aromatic	Colored
6. Miscellaneous alkaloids	Silica gel G	a. $CHCl_3$: acetone: diethylamine (5:4:1) b. $CHCl_3$: diethyl-amine (9:1) c. Cyclohexane: $CHCl_3$: diethyl-amine (5:4:1) d. Cyclohexane: diethylamine (9:1)	Dragendorff's reagent

Table 3.4, continued

Compounds	Adsorbent	Developer	Visualization
	Aluminum oxide	e. Toluene: EtOAc: diethylamine (7:2:1)	
		a. $CHCl_3$	
		b. Cyclohexane: $CHCl_3$ (3:7) plus 0.05% diethylamine	
7. Alkaloids and barbiturates in toxicology	Silica gel G	a. MeOH b. $CHCl_3$: EtOEt (85:15)	Dragendorff's reagent
8. Strong-base amines	Unbound aluminum oxide	a. Acetone: heptane (1:1) b. $CHCl_3/NH_3$ (sat. at 22°): EtOH (96%) (30:1)	I_2 vapor – U.V. (Dragendorff's reagent)
9. Sugar acetates and inositol acetates	Silica gel G	Toluene with 2–10% MeOH	H_2O

Table 3.4, continued

Compounds	Adsorbent	Developer	Visualization
10. Aldehyde 2,4-DNPH's	a. Aluminum oxide G b. Silica gel G	a. Toluene: EtOAc (1:1) b. Toluene: EtOAc (3:1) (1:1)	a. Colored compounds b. NaOH
11. Dicarboxylic acids	Silica gel G	a. Toluene: MeOH: HOAc (45:8:4) b. Toluene: dioxane: HOAc (90:25:4)	Bromphenol blue acidified with citric acid
12. p-Hydroxybenzoic acid esters	Silica gel G	Pentane: HOAc (88:12)	(U.V. on phosphors)
13. Sulfonamides	Silica gel G	CHCl$_3$: EtOH: heptane (1:1:1)	p-Dimethylamino-benzaldehyde, acidified (U.V. on phosphors)
14. Food dyes	Silica gel G	a. CHCl$_3$: acetic anhydride (75:2) b. Toluene c. Methyl ethyl ketone: HOAc: MeOH (40:5:5)	SbCl$_3$ in CHCl$_3$

Table 3.4, continued

Compounds	Adsorbent	Developer	Visualization
15. Miscellaneous essential oils	Silica gel G	Toluene: CHCl$_3$ (1:1)	SbCl$_3$ in CHCl$_3$
16. Alkali metals: Na$^+$, Li$^+$, K$^+$, Mg$^+$	Purified silica gel G	EtOH: HOAc (100:1)	Acid violet (1.5% aqueous soln.)
17. Ferrocene derivatives	Silica gel G	a. Toluene b. Toluene: EtOH (30:1) (15:1) c. Propylene glycol: MeOH (1:1) d. Propylene glycol: chlorobenzene: MeOH (1:1:1)	(U.V. on phosphors)
18. Fatty acid methyl esters	Silica gel G	Hexane: EtOEt mixtures (up to 30% EtOEt)	a. I$_2$ b. H$_2$SO$_4$ and heat
19. Glycerides	Silica gel G	a. Skellysolve F: EtOEt (70:30) b. (10:90) for monoglycerides	H$_2$SO$_4$ and heat

Table 3.4, continued

Compounds	Adsorbent	Developer	Visualization
		c. (70:30) for diglycerides d. (90:10) for triglycerides e. (60:40) (35:65) (85:15)	
20. Long chain aliphatic alcohols	Silica gel G	a. Pet ether: EtOEt: HOAc (90:10:1)	H_2SO_4 and heat
21. Phenols	Plaster of Paris bound silica gel	a. Hexane: EtOAc (4:1) (3:2) b. Toluene: EtOEt (4:1) c. Toluene	(U.V. on phosphors)
22. 3,5-Dinitrobenzoates of alcohols and phenols	Silica gel G	a. Toluene: pet ether (1:1) b. Hexane: EtOAc (85:15) (75:25) c. Toluene: EtOAc (90:10)	Colored compounds

Table 3.4, continued

Compounds	Adsorbent	Developer	Visualization
23. Steroids	Silica gel G	a. Toluene b. Toluene: EtOAc (9:1) (2:1) c. Cyclohexane: EtOAc (9:1) (19:1) d. 1, 2-dichloroethane	$SbCl_3$ in $CHCl_3$
24. Plasticizers (phthalates, phosphates, and other esters)	Silica gel G with phosphors	a. Isooctane: EtOAc (90:10) b. Toluene: EtOAc (95:5) c. Butyl ether: hexane (80:20)	a. U.V. b. I_2

[a] Directions for the preparation of these reagents are found in Table 4.6.

Table 3.5   SOLVENTS FOR COLUMN PARTITION
            CHROMATOGRAPHY

Stationary Phase	Mobile Phase
Normal Partition	
Water	Alcohols (n-butanol, isobutanol)
Water plus acid	
Water plus alkali	Hydrocarbons (benzene, toluene, cyclohexane, hexane)
Water plus buffer components	Chloroform
Aqueous alcohols (MeOH, EtOH)	Ethyl acetate
Alcohols (MeOH, EtOH)	Ethylene glycol monomethyl ether
Formamide	Methyl/ethyl ketone
Glycols (ethylene, propylene, glycerol)	Pyridine
Reversed Phase Partition	
n-Butanol	Water
Octanol	Water plus acid
Chloroform	Water plus alkali
Chlorosilanes and silicones	Water plus buffer components
Mineral oil	Aqueous alcohols (MeOH, EtOH)
Paraffin	Alcohols (MeOH, EtOH)
	Formamide
	Glycols (ethylene, propylene, glycerol)

Prepare or purchase some thin layers of silica gel, cellulose, or kieselguhr and place them in a dessicator or similar container over water for 12-24 hr to saturate them.   Spot the layers with the sample to be investigated and develop them with a series of liquids prepared in the following manner.   Shake *n*-butanol with water in a separatory funnel, allow the layers to separate, and use the *n*-butanol layer.   This procedure produces *n*-butanol saturated with water (the liquid stationary phase).   In a similar manner the other three liquids can be prepared from *n*-butanol and 1N ammonium hydroxide, *n*-butanol and 1N aqueous acetic acid and ethyl acetate and water.   Of these four liquids, two are neutral, one is acidic, and one is basic.   Visualize the chromatograms and compare.   The more promising systems can then be modified to produce optimum results.   These modifications can be made by altering the polarity of the organic solvent with other solvents or by altering the acidity or basicity of the stationary phase liquids.

A number of layers containing stationary phases similar to the highly specialized packings used in HPLC (Chapter 6) are available commercially and should be used to scout such systems.   As stated previously, paper chromatography (Section 3.7) is, in essence, a type of TLC using an LLC system, and can be used to scout such methods.

Several LLC systems that have been worked out for column LLC chromatography are given in Table 3.6.   These have been abstracted from the book by Cassidy.[3]   In the last columns in this table, we have suggested some visualization methods that would be suitable for use in TLC.   The abbreviations are those used in Table 3.3.

REFERENCES

[1]R. Neher and E. von Arx, *Steroid Chromatography,* American Elsevier Publishing Co., Inc. **(1964)** p. 249.

[2]J.M. Bobbitt, *Thin Layer Chromatography,* Reinhold Book Corporation, New York **(1963)** pp. 128-182.

[3]H.G. Cassidy, *Fundamentals of Chromatography,* Interscience Publishers, Inc., New York **(1957)** pp. 126-130.   Original literature is cited therein.

Table 3.6   COLUMN–PARTITION SYSTEMS

Compounds	Stationary Phase	Moving Phase[a]	Visualization (TLC)[b]
On Silica Gel			
1. n-Acetylpeptides	H_2O	a. n-BuOH : $CHCl_3$ b. EtOAc : H_2O	H_2SO_4 and heat
2. C_2–C_{12} Aliphatic acids	NaOH–MeOH (7.5 mL of 1 N NaOH made up to 1 with MeOH)	a. Isooctane:EtOEt (1:9) b. EtOEt	H_2SO_4 and heat
3. C_4–C_6 Dicarboxylic acids	H_2O	n-BuOH: $CHCl_3$ (1:9) followed by (1:4)	Bromcresol green
4. Aromatic di– and tribasic acids	H_2O	n-BuOH: $CHCl_3$ solutions (increasing n-BuOH by gradient elution)	Bromcresol green
5. Monohydric alcohols	H_2O	CCl_4 with increasing amounts of $CHCl_3$ and finally $CHCl_3$: HOAc (9:1)	H_2SO_4 and heat

Table 3.6,   continued

Compounds	Stationary Phase	Moving Phase[a]	Visualization (TLC)[b]
6. Aldehydes as semicarbazones	H_2O	$CHCl_3 : n$-BuOH	H_2SO_4 and heat on silica gel or kieselguhr
7. 2,4-Dinitrophenyl amino acids	H_2O	n-BuOH : $CHCl_3$	Colored compounds
8. Partially methylated glucoses	H_2O	n-BuOH : $CHCl_3$	H_2SO_4 and heat
9. Phenols and cresols	H_2O	Cyclohexane	U.V. on phosphors
On Kieselguhr (Diatomaceous Earth)			
1. C_2–C_{10} Aliphatic acids	27–35 N H_2SO_4	a. Toluene b. Toluene: pet ether	Heat
2. Di- and trihydric alcohols	H_2O	a. EtOAc b. Toluene: n-BuOH	H_2SO_4 and heat
3. Penicillins	pH 5.5 Citrate buffer	EtOEt: diisopropyl ether (1:1)	H_2SO_4 and heat

Table 3.6, continued

Compounds	Stationary Phase	Moving Phase[a]	Visualization (TLC)[b]
4. Pentose nucleosides and nucleic acids	H_2O	n–BuOH	U.V. on phosphors
On Cellulose			
1. Amino acids	H_2O	a. n–BuOH b. n–BuOH: HOAc: H_2O (3:1:1) c. Phenol: H_2O (3:1)	Ninhydrin
2. Sugars and derivatives	H_2O	n–BuOH	Aniline phthalate or anisaldehyde
3. Methylated sugars	H_2O	Ligroin: n–BuOH (3:2)	Anisaldehyde

[a] It should be understood that the moving phase has been saturated with the stationary phase by shaking the two together in a separatory funnel.

[b] Directions for the preparation of these reagents are found in Table 4.6.

Chapter 4

THIN LAYER
AND PAPER CHROMATOGRAPHY

4.1 INTRODUCTION

Thin layer chromatography (TLC) and paper chromatography (PC) are the simplest of the liquid chromatographic methods that will be presented in this book. Since PC has been largely displaced by TLC in most laboratories, we will discuss TLC almost entirely. A brief section at the end of the chapter will be used to summarize PC and contrast it with TLC.

The basic steps of TLC and PC have been described in Chapter 1, and TLC was illustrated in Figures 1.4-1.7. Using TLC, separations of such widely differing substances as naturally occurring or synthetic organic compounds, inorganic-organic complexes, and even inorganic ions can be achieved in a few minutes with equipment costing only a few dollars. Quantities as low as a few micrograms or as large as 5 g can be dealt with, depending on the equipment available and the chromatographic phenomena involved. Other advantages of TLC are the use of small amounts of solvents and samples, the possibility of **multiple sample spotting** (making direct comparisons between samples practical), and the wide range of methods available (such as LSC, LLC, and exclusion chromatography).

An extensive literature now exists on TLC separations. This is summarized in some of the books listed in the Bibliography in the back of the book and in the *Journal of Chromatography*. The data are given in terms of Rf values of specific compounds in certain solvent systems. Rf was defined in Chapter 1 and can be related to the other theoretical concepts by equation 1.3. An important series of four papers by Guiochon[1] describes his study of TLC, and includes a discussion of the optimization of experimental conditions.

TLC may be used with two goals in mind. First, it may be used in its own right as a method for achieving qualitative, quantitative, or preparative results (see discussion in Chapter 1). Second, TLC may be used to explore solvent systems and support systems to be used in column chromatography (Chapter 5) or high performance liquid chromatography (Chapter 6), as mentioned in Chapter 3 and summarized in Tables 3.3 and 3.4. These methods will be specifically described in the later chapters.

Probably the most remarkable development in TLC that has taken place in recent years has been the widespread adoption of commercially prepared layers, as opposed to layers prepared by the user. Layers are available consisting of almost any adsorbent or support, of almost any thickness, and in almost any size. They are available on glass backing, plastic backing, and on metal foil. The commercial sources are summarized in Table 4.3. Although the chapter will be presented in a do-it-yourself fashion (some of us are still old-fashioned), it should be understood that commercial layers are available for any use described. Still another recent development has been HPTLC or high performance thin layer chromatography using special adsorbents and techniques. This will be considered in Section 4.7.

Like most techniques, TLC can be used on several levels of complexity or sophistication. These are, in increasing complexity: microscope slide TLC, large layer TLC, preparative TLC, quantitative TLC, and HPTLC. The material in this chapter will be organized and presented along these lines.

TLC on microscope slides (or commercial layers cut to smaller sizes) yields separations of mixtures of up to four components in about 5 min using normal laboratory glassware. The layers are easy to prepare, generally require no activation, and yield sharp separations. For these reasons, the method is preferred by many investigators and is an ideal method for use in undergraduate laboratories.

TLC on larger layers, generally 5 x 20 cm, 10 x 20 cm, or 20 x 20 cm, requires either their purchase or the special equipment needed for their preparation. Development times of 30 min to an hour are normal. However, they may be used to separate a greater number of mixtures containing a larger number of components and are amenable to special techniques of development such as two-dimensional TLC, TLC on shaped layers, and continuous TLC. Preparative and quantitative TLC are usually carried out on the larger layers.

TLC involves essentially two variables: the nature of the stationary phase or layer and the nature of the mobile phase or development solvent mixture. The stationary phase can be a finely divided powder functioning as an adsorbing surface (liquid-solid chromatography) or as a support for a liquid film (liquid-liquid chromatography). (The stationary phase in TLC is frequently called the adsorbent, even when it is functioning as a support for a liquid in a LLC system.) Almost any powder can be and has been used as an adsorbent in TLC, but we shall confine our discussion to the four most commonly used: silica gel (silicic acid), alumina (aluminum oxide), kieselguhr (diatomaceous earth), and cellulose. The mobile phase can be almost any solvent or mixture of solvents. The choice of mobile phases was discussed in detail in Section 3.7.

In the following sections the levels of complexity mentioned above will be discussed. Section 4.2 will contain a complete discussion of TLC on microscope slides and should suffice for work at that level. Section 4.3 will include a discussion of TLC on larger layers which, when taken with Section 4.2, will be complete. Sections 4.4 and 4.5 will be devoted to the additional complexities of preparative and quantitative TLC; Section 4.6 will involve special problems in TLC; Section 4.7 will be devoted to HPTLC; and Section 4.8 will be concerned with paper chromatography.

4.2 CHROMATOGRAPHY ON MICROSCOPE SLIDES OR SMALL LAYERS

Much of the information in this section is taken from a remarkable paper by J. J. Peifer[2] of the Hormel Institute in Austin, Minnesota. TLC on microscope slides represents the simplest of the many systems that have been proposed, and we will use it to present a simple but complete picture of TLC.

Adsorbents and Additives

Adsorbents for TLC are, in order of importance, silica gel, alumina, kieselguhr, and cellulose. They are more finely divided (passing a 200-mesh screen) than those adsorbents used in classical column chromatography and are on the order of those used in HPLC. In fact, TLC adsorbents can be used directly in some HPLC systems. The adsorbents usually contain a binder and many also contain a number of other additives (see below).

There is a very appreciable difference between adsorbents (and prepared layers) from different commercial sources. Silica gel, in particular, has a wide range of properties, sometimes even differing from one batch to another of the same manufacturer. It is advisable to work, as much as possible, with the product of one company. Although directions are available in the literature for the preparation of the various adsorbents, it is recommended that they be purchased from commercial sources (see Suppliers).

Silica Gel. Silica gel is the most extensively used adsorbent in TLC and HPLC and would be the logical material to try first. Neutral compounds containing up to three functional groups should be resolvable on activated layers using normal organic solvents or mixtures of solvents. Since most silica gels are slightly acidic in nature, acids are frequently fairly easily separated, thus minimizing acid-base reactions between the adsorbent and the material being separated. If difficulties are encountered in separating acids or bases, it may be necessary to use acidified or basified solvents, as described in the chapter on solvent choice, in Section 3.6.

Alumina. Alumina, in contrast to silica gel, is slightly basic in nature and is often used for the separation of bases. This technique minimizes acid-base reactions also. TLC on alumina is often used as a quick, qualitative method for predicting solvent systems in column chromatography in which alumina has been more consistently used than silica gel.

Kieselguhr and **Cellulose.** Both kieselguhr and cellulose are support materials for a liquid film used in an LLC system, and thin layers of cellulose are closely related to classic paper chromatography. These types of chromatography are always used for the separation of such highly polar compounds as amino acids, carbohydrates, nucleotides, and various other naturally occurring hydrophilic compounds. Since LLC has a higher degree of possible resolution (Chapter 3), it is sometimes used for the separation of closely related isomers. Care must be taken to see that the layers do, in fact, contain a film of liquid (see below).

Water. The presence or absence of water in chromatographic adsorbents or supports is extremely important. Layers of silica gel or alumina that will be used for adsorption work (LSC) must contain a minimum of water, otherwise water will occupy all of the adsorption sites and no attachment of solute can take place. Layers with the minimum amount of water are said to be **activated** and are prepared by heating at 100°C or so for 1-3 hr. If activation temperatures are much over 110°C, irreversible dehydrations may take place on the adsorbent, leading to less effective separations. The layers should then be stored in a dessicator or dry box. Commercial layers do vary in their activation, but can usually be used as such, or can be further activated with heat.

On the other hand, kieselguhr or cellulose layers or even silica layers that are to be used for LLC *must* contain a film of liquid, that is usually water. If this liquid is not present from the mode of formation and if water is to be used, the layers must be hydrated.

Binders. Thin layers are held together and on a glass plate by various binders. The most common of these, at least for homemade layers, is plaster of Paris (hydrated calcium sulfate), that is added to adsorbents in amounts up to 10-15%. Most commercial layers are held together with an organic polymer such as polyvinyl alcohol and are harder and more stable than homemade layers. Other binders that have been used are starch (in amounts up to 3%) and low molecular weight silicates. Layers can also be prepared, and are commercially available, that contain no binder but are held together by virtue of a narrow range of particle sizes in the adsorbent.

Visualizing Agents. Thin layers often contain a phosphor which is added to aid the visualization of colorless spots on the developed layers. Phosphors are substances that emit visual light when irradiated with light of another wavelength, generally in the ultraviolet region. Thus, a layer containing a phosphor will shine (like a TV screen) when irradiated at the correct wavelength. If the compound to be visualized in a spot contains conjugated double bonds or an aromatic ring of any type, the exciting UV light cannot reach the phosphor, and no light can be emitted. The result is a dark spot on a shining background. The method is quite sensitive and does not destroy the compound being visualized. The most useful phosphors are inorganic sulfides that emit light when irradiated at 254 nm.

Phosphors are present in commercial adsorbents and layers in amounts of about 1% and appear to play no part in the chromatographic process. Organic phosphors that emit light when irradiated at 360 nm are also available, but are less desirable because they are organic compounds and sometimes move in the chromatographic system. Some organic compounds shine or fluoresce themselves when irradiated at 254 or 360 nm and can be easily visualized.

Layer Preparation

Although we will describe the preparation of microscope slide layers, it should be noted that such layers are also commercially available. Layers on plastic or foil backing can be cut to any size with scissors and glass plates can be cut to size by scoring the back with a glass cutter and then breaking the glass plate. Glass backed layers are even available with pre-scored lines to make them easy to break up into small layers.

Thin layers are most conveniently prepared on microscope slides by dipping the slides (two at a time) into a slurry of adsorbent in a mixture of organic solvents. Directions for these slurries are given in Table 4.1. There will be some variation in the solid:solvent ratios depending upon the source and type of adsorbent. Within limits, thicker slurries will yield thicker layers. Thin and grainy layers result when the slurry is too thin. The slurries are stable for several weeks when kept in a well sealed container.

The following procedure should produce satisfactory layers:

1. Prepare a slurry as described in Table 4.1. Shake for about 2 min.

2. Dip two microscope slides, held back to back into the slurry as shown in Figure 4.1. Withdraw them slowly and allow them to drain on the edge of the container.

3. Separate the slides and wipe the excess adsorbent off the edges. Allow the plates to dry for 5-20 min, preferably in a hood. Layers of silica gel and alumina prepared in this fashion are reasonably, but not completely, activated. They should be prepared anew each day or kept in a dry atmosphere.

4. When the layers are to be used in partition chromatography with water as a stationary phase (on silica gel, cellulose, or kieselguhr), they must be rehydrated before use, since they have been prepared in an essentially dry system. This can be done by holding them over a beaker of boiling water and allowing them to dry in the atmosphere at room temperature.

5. After use, the microscope slides should be wiped off, washed with soap or detergent, rinsed carefully with water, and dried for reuse.

Table 4.1   RECIPES FOR THE PREPARATION OF PEIFER SLURRIES

Adsorbent	Slurry Medium	Proportions, g in mL
Silica gel G[a]	Methylene chloride: methanol (2:1, v/v)	35 g in 100 mL
Cellulose powder[b]	Methylene chloride: methanol (50:50, v/v)	50 g in 100 mL[b]
Alumina[c]	Methylene chloride: methanol (70:30, v/v)	60 g in 100 mL[c]

[a] Any of the commercial plaster of Paris bound adsorbents may be substituted. The adsorbents may, if desired, contain phosphors.

[b] Most cellulose powers will form satisfactory layers without plaster of Paris due to their fibrous natures. However, especially good layers can be prepared by triturating 35 g of cellulose and 15 g of plaster of Paris in a minimum of methanol and diluting the viscous paste to give the above ratio.

[c] 45 g of activated alumina and 15 g of plaster of Paris (or 60 g of commercial bound adsorbent containing this binder) are triturated with a minimum volume of methylene chloride: methanol and diluted to the above ratio.

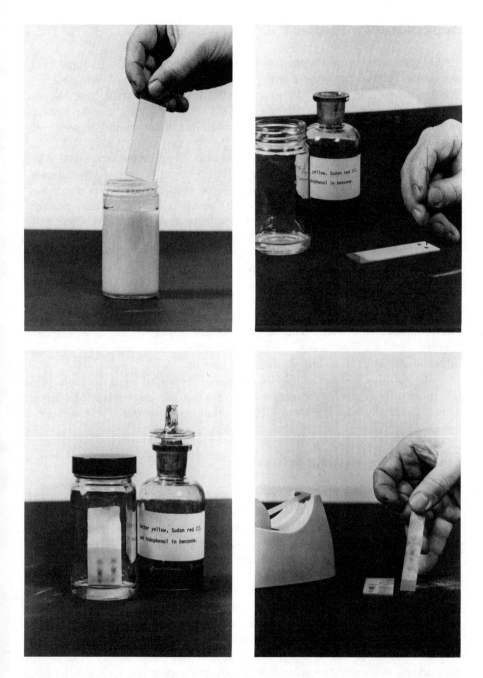

Figures 4.1–4.4 Two microscope slides, back to back, are dipped in a silica gel slurry (4.1), separated, and allowed to dry. They are spotted (4.2) with a dye mixture and developed with benzene (4.3). The cellulose tape impression is taken from the dried, developed layer (4.4) for recording in a laboratory notebook.

Application of the Sample

The mixture to be chromatographed should be dissolved in some fairly non-polar solvent for application (spotting) onto the layer. In general, a 0.1-1% solution is used. Almost any solvent will do, but the best solvents are those boiling between 50 and 100°C. Such solvents can be handled easily and will evaporate from the layer readily. Water should be used only when there is no other choice.

The two major disadvantages of TLC on coated microscope slides are that the layers are relatively thin compared to larger homemade or commercial layers and that the distance available for chromatography is much less. Thus, it becomes necessary to apply the samples in the smallest possible area. This spotting can be done with a fine capillary pulled from glass tubing such that it is not much thicker than a straight pin. The sample, in solution, should be applied about 8-10 mm from the end of the slide, which is completely coated (Figure 4.2). Several applications can be made to the same spot as long as the layer is dried between applications, and as many as three sample spots can be placed on one layer if proper care is taken. If only one sample is being chromatographed, it is advisable to spot it in three different concentrations. The spots from the smallest sample will appear sharper and with less tailing. The spots from the larger samples will reveal information about trace impurities in the sample.

The spotting solvent should be completely removed from the layer prior to chromatography, if necessary with a hot air gun or electric hair dryer. This is especially important and difficult when water is used.

Choice of a Development Solvent

Solvent systems for TLC can be chosen from the literature, but more frequently one chooses by trial and error, since the time required for such experiments is short. The simplest systems are mixtures of organic solvents used to separate mono- and difunctional molecules by LSC on activated silica gel or alumina layers. The choice of a solvent system was discussed in a general way in Chapter 3.

A fairly systematic attempt to find a suitable development solvent can be made using Table 3.1. Four solvents are underlined in the table: light petroleum ether, toluene, ethyl ether, and methanol, ranging from the least to the most polar. A chromatogram should be run in each of the four solvents. In some cases, a pure solvent will yield satisfactory results. In most cases, however, one solvent will move the spots too far, and the next one up the list (less polar) will not move them far enough. One should then mix solvents to obtain a desired polarity. Cautions about mixing solvents are discussed in Chapter 3 and should be reviewed.

Development of Microscope Slide Layers

The spotted layer is placed in a small jar containing a layer of solvent a few millimeters deep, as shown in Figure 4.3. The solvent level in the jar should be below the spots on the plate. The jar is closed with a lid or piece of aluminum foil, and the solvent is allowed to ascend about three-fourths of the way up the layer. Development will require about 5 min, depending upon the adsorbent and the solvent. If the spots have an Rf of less than 0.5, the layers should be dried and rechromatographed. This technique, known as **multiple development,** will always improve a separation and will be discussed in detail later in this chapter.

Visualization of Microscope Slide Layers

When all of the compounds being chromatographed are colored, it is easy to see whether and how well they are separated. When some or all the compounds are colorless, as is usually the case, they must be visualized by some method or reagent. Visualization techniques are either specific methods (devised to show specific functional groups of types of compounds) or general methods that show any organic compound. General methods used on small layers are the adsorption of iodine vapor, the use of UV light, and the use of UV light on phosphor containing layers. The UV phosphor method was discussed above under adsorbent additives.

Visualization with iodine is carried out by placing the developed and dried chromatogram in a jar containing iodine crystals. After the jar is closed, the iodine vapor will adsorb slowly into the spots on the layer that contain organic compounds, and the spots will show up as brown areas on a white background. The method may require 5-10 min and is fairly general for organic compounds. The iodine will sublime away after the layer is removed from the jar, and the spots will slowly fade. The method is nondestructive in most cases; that is, the organic compounds in the spots are not destroyed and can often be isolated.

Specific visualization methods and sulfuric acid charring are more commonly used on larger and more stable layers and will be discussed below.

Documentation

Microscope slide-sized chromatograms can best be recorded in a laboratory book or report by attaching the layer to transparent tape. The tape is pressed into the layer as shown in Figure 4.4. The tape and the portion of the layer adhering to it are removed, and an additional layer of tape is placed on the back, covering the layer. The resultant "sandwich" is then taped into a laboratory book. When plastic or foil backed commercial layers are used, they can be placed directly into a book after the layer has been covered with tape. One can, of course, always draw a picture of the chromatogram or record the Rf's of the components.

4.3 TLC ON LARGER LAYERS

Macro-layer TLC is generally carried out on two standard sizes: 5 x 20 cm and 20 x 20 cm. In theory, of course, any size layers can be used, but the problem that arises with very large layers, for example, 20 x 1000 cm, is that one must have very large and expensive development chambers. Larger layers have several advantages over the smaller layers. They are usually thicker, adhere to the glass plate a little better, and provide a larger area for chromatography. It is possible to investigate more samples simultaneously (generally four on the 5 x 20 cm layers and 18 on the 20 x 20 cm layers), and the development track is longer, permitting the resolution of more complex mixtures. A number of special development techniques are better carried out on larger layers; some HPTLC methods require the larger layers.

Larger layers of many sizes and types are available commercially, as shown in Table 4.2. (The F means that a phosphor is present.) In general, they are harder, more uniform, and more expensive than homemade layers. Large glass plates can be purchased pre-scored so that they can be easily broken to yield 5 x 20 cm layers, and layers on plastic backing can be cut with scissors to any desired size. In addition, some glass plates are pre-channeled and/or pre-numbered to permit a large number of samples to be run both efficiently and effectively. The commercial layers are used by most industrial laboratories, by many academic research groups, and by some laboratory classes. They overcome some of the disadvantages of homemade layers.

Homemade layers have several disadvantages and generally require some type of commercial equipment for their preparation. Since the layers are cast from an aqueous slurry, they must be heated for at least 1 hr at 110°C to activate them for use in LSC. Once activated, they must be stored in a dry atmosphere, thus requiring large dessicators or special boxes of some type.

Much commercial equipment has been developed to prepare and deal with larger TLC layers. Some of this equipment will be discussed in the following sections, but no attempt will be made to mention everything available. The list of supplies and manufacturers in the Suppliers section should be relatively complete.

It should be understood that the material in this section represents an extension of Section 4.2 and that a knowledge of that section is assumed.

Table 4.2   COMMERCIALLY AVAILABLE THIN LAYERS
FOR CHROMATOGRAPHY

Adsorbent	Type
Silica gel G, Silica gel GF	Normal Pre-channeled[a] Pre-scored[b] Pre-adsorbent[c] Pre-numbered Pre-saturated Hard layer High performance
Alumina G, Alumina GF	Normal Pre-scored
Cellulose, Microcellulose (Avicel and Avicel F)	Normal Pre-scored

[a] *Channels in TLC layer.*

[b] *Glass pre-scored to permit facile breaking.*

[c] *Spotting area composition concentrates sample before it contacts the silica, leading to more discrete and reproducible "bar"-shaped samples.*

Adsorbents

As in Section 4.2, the discussion will be limited to four materials: silica gel, alumina, kieselguhr, and cellulose.

Commercial Adsorbents. Commercial adsorbents for TLC are available in a confusing array. Table 4.3 lists many of them. Adsorbents do differ appreciably from one manufacturer to another in such properties as pH, pore size, flow rates, purities, and powers of resolution. The properties of some of the binders and phosphors present in adsorbents were discussed above.

Cleaning Adsorbents. Commercial adsorbents, and commercial layers as well, often contain a number of impurities that may or may not cause

Table 4.3   COMMERCIAL TLC ADSORBENTS

Adsorbents

Acetylated cellulose	Silica – C-18 bonded phase
Alumina G	Silica gel G^a
Alumina GF	Silica gel GF^b
Alumina H	Silica gel H^c
Alumina HF	Silica gel G/AgNO$_3$
Cellulose	Silica gel GF/AgNO$_3$
Florisil	Woelm alumina acidic
Florisil F	Woelm alumina acidic F
Kieselguhr	Woelm alumina basic
Kieselguhr F	Woelm alumina basic F
Microcellulose	Woelm alumina neutral
Microcellulose F	Woelm alumina neutral F
Polyamide	
Sephadex	

Sources For Layers

Ace Scientific	Gelman Instrument Co.
Alltech Associates	ICN Pharmaceuticals, Inc.
Analtech	Kontes Glass Co.
Analytichem, Inc.	Pharmacia Fine Chemicals
J. T. Baker	Quantum Industries
Bio-Rad	Regis Chemical
Brinkmann/Sybron	Schleicher & Schuell
Camag	Supelco
EM Laboratories, Inc.	A. H. Thomas
Eastman Kodak	Whatman, Inc.

^{a} *Contains gypsum (plaster of Paris).*

^{b} *Contains inorganic fluorescent phosphor.*

^{c} *Contains fine silica.*

problems.   The impurities generally fall into three groups: inorganic ions, low molecular weight silicates (in silica gel), and miscellaneous organic compounds.

Low molecular weight silicates and organic impurities can be removed by pre-washing the adsorbent in bulk with methanol.   This is important when one wants to remove the separated compounds for quantitative TLC or isolation in preparative TLC.   The purification is carried out by boiling a slurry of adsorbent in methanol for a few minutes and allowing it to stand for a few hours at room temperature. The adsorbent should then be collected by filtration, dried at room temperature and then at 110°C for 1 hr. The adsorbent should still contain sufficient binder (when it is plaster of Paris) or phosphor (when it is the inorganic type that fluoresces at 254 nm).

Commercial or homemade layers can be cleaned by pre-development with a solvent that is more polar than the one that will be used in the final chromatography. For example, one can develop the layers with methanol to drive the impurities to one edge of the layer. The layer can then be reactivated, and the cleaned area can be used for chromatography.

Preparation of Layers

Supporting Plates. Large thin layers are usually prepared by spreading a film of an adsorbent-water slurry onto a support surface and allowing the slurry to dry. The supporting surface is generally a piece of plate glass with smoothed edges (to prevent cuts). Pyrex glass, stainless steel, aluminum, and many other materials have been used for this purpose, but appear to offer little advantage. Commercially prepared layers, are also backed with plastic films or aluminum foil.

The glass plates should be washed with detergent, rinsed well, dried, and wiped with a wad of clean cotton soaked with hexane before the slurry is cast. The hexane removes any traces of oils and thus produces a better adherence between the layer and the plate.

Slurries. The optimum thickness of a slurry used to cast a thin layer will depend, within limits, on the method used to cast the film. If the slurry is too thin, it will run off the plate; if it is too thick, it will not flow out of an applicator. Most adsorbent manufacturers suggest adsorbent:water ratios, some of which are given in Table 4.4. The adsorbent and water are either shaken together in a stoppered Erlenmeyer flask, ground together with a mortar and pestle, or slurried in a blender (for cellulose layers). The slurry, especially when it contains plaster of Paris as a binder, becomes thicker with time, and a uniform shaking time (*ca.* 45 sec) is suggested. A pea soup or milkshake consistency is about right for the slurry. One should not hesitate to deviate from the ratios given in Table 4.4 when this seems desirable. The precise amount of water needed often depends on the type and previous history of the adsorbent.

Layer Casting. Layers of almost any size and up to a thickness of 1 mm or so can be prepared with a glass plate, some masking tape, and a glass rod.[3] Layers of masking tape (up to five layers) are built up on opposite sides of a glass plate. The slurry, perhaps a little thicker than usual, is poured on the plate and leveled with a glass rod that is supported on the tape. The thickness of the tape determines the layer thickness and

Table 4.4   RECOMMENDED AMOUNT OF WATER TO BE
ADDED TO VARIOUS COMMERCIAL ADSORBENTS

Adsorbent	Adsorbent:Water (g:mL)
Silica gel G	30:60–65
Silica gel H	30:80–90
Alumina G	30:40
Alumina H	30:80–90
Cellulose	15–20:60–90
Kieselguhr	30:60–65
Polyamide	15:65
Sephadex	4.5–10.5:100

must be removed before activation. The process, although simple and inexpensive, is not efficient for the preparation of larger numbers of layers, but can be used for layers larger than 20 x 20 cm.

A large amount of equipment is commercially available for the preparation of thin layers. There are essentially two basic designs. In one design, the glass plates are laid out on a flat surface or form and an applicator is drawn over them depositing a film of slurry. In the second design, the glass plates are pushed under a hopper that deposits the slurry.

The Stahl-Desaga apparatus (Brinkmann/Sybron) is an example of the first type and is shown in Figure 4.5. The glass plates may be 5 x 20 cm, 10 x 20 cm, or 20 x 20 cm and are held in a plastic holder. The applicator is available in several models, of which the most useful is the variable thickness design. Layers may be prepared in thicknesses up to 2 mm. The system is convenient and efficient, but requires that all plates have the same thickness. Otherwise, one obtains layers of varying thicknesses. Apparatus of this general type is available from other companies mentioned in the Suppliers section in back of the book.

Modified Layers. A number of substances can be incorporated into thin layers for specific purposes. In general, these substances are acids, bases, buffers, and complexing agents.

In Section 4.2 we discussed the pH differences between adsorbents, especially between slightly acidic silica gel and slightly basic alumina, and the role that these differences might play in the separation of organic acids

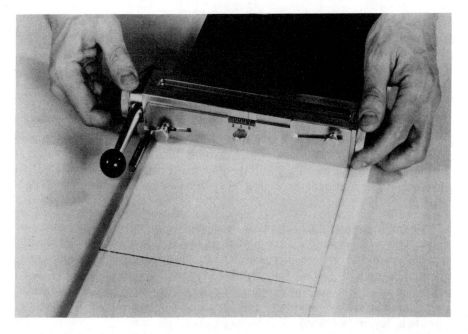

Figure 4.5 The Stahl-Desaga apparatus in use, showing the adjustable feature of the spreader.   (Reproduced through the courtesy of Brinkmann/Sybron Instruments, Inc.)

and bases. It is possible to modify the pH of any layer by using acids, bases, or buffers rather than pure water for adsorbent slurries. Thus, one might use 0.5N oxalic acid or 0.5N sulfuric acid to prepare layers that would be used to separate acids and neutral compounds while holding bases at the origin. Layers prepared with 0.5N sodium hydroxide can be used for the separation of bases. Such acidic or basic layers help to prevent the tailing that may result when acids or bases are chromatographed on a neutral layer. Buffers of all types may be used. It should be noted that the nature of a slurry and its rate of thickening may be changed by these additives, and it is not certain whether these layers function by LSC or LLC or a mixture of the two. They do, however, work.

The addition of complexing agents to thin layers can alleviate two problems in TLC. The first of these involves the separation of very nonpolar compounds such as alkanes, alkenes, alkynes, and aromatic hydrocarbons. These materials have little attraction to even the most activated layers and are moved with the **solvent front** by the least polar solvent, hexane. When the layer contains a complexing agent such as silver nitrate (12.5% in the slurry), it will have an enhanced tendency to hold and separate alkenes and aromatic hydrocarbons, which form complexes with the silver ion. Furthermore, complex formation between double bonds and silver ion is a function of the stereochemistry of the alkene, so that silver-impregnated layers can be used to separate unsaturated isomers. Slurries containing silver nitrate should not be allowed to stand in stainless steel applicators such as the Stahl-Desaga spreader.[4]

The second problem involves the separation of mixtures of very polar compounds such as the sugars. In this case, one can use the complex formation reaction between diols and borate ion to prepare a more discriminating layer. The slurries are made with 0.1N boric acid solutions.[5]

Layer Activation. The cast slurry films of silica gel or alumina should be allowed to stand for 30 min or so at room temperature and activated at 110°C for at least 1 hr. They will then have a Brockmann[6] activity of II-III (Chapter 5) and should be stored in a dry-box or large dessicator until used.

Layer Impregnation. Methods for placing a liquid stationary phase on a thin layer will depend on the nature of the stationary phase. When the liquid is to be water, the layers cast from aqueous slurries are allowed to dry and heated to 110°C for 10 min. When the slurries contain added acids, bases, or buffers, the resulting stationary liquid phase is also acidic, basic, or buffered. Such systems account for most LLC. The pre-adsorbent layers (see Table 4.2) facilitate this technique.

Various other liquids can, of course, be used as stationary phases (Table 3.5). They may be polar liquids such as formamide or ethylene glycol or non-polar liquids such as silicone oil or paraffin oil. In general, layers are activated to remove the water from them and then impregnated with the desired liquid. The liquids may be conveniently added by dipping an activated layer (silica gel or kieselguhr) slowly into a solution of the liquid in a volatile solvent 20% formamide or ethylene glycol in acetone, 5% silicone oil (Dow-Corning 200, 10 cs viscosity) in ether or 15% undecane in hexane. The layers are allowed to dry without heat.

Sample Preparation and Application

Samples for chromatography are applied as solutions of the mixture to be separated. This has been discussed previously. In general, amounts on the order of 50-100 μg are applied to single spots for adsorption chromatography and 5-20 μg are applied for partition chromatography.

It is possible to spot certain types of compounds as their salts if a suitable reagent is placed in the development solvent to regenerate them. For example, amine hydrochlorides can be spotted and will be chromatographed as free bases if a small amount (about 0.1%) of ammonia or diethylamine is added to the development solvent. The precision needed for the application of solution to the layer depends on the goals of the chromatography. For qualitative TLC, it is probably more important to investigate the mixture at several relative concentrations than to investigate a precisely known amount of sample. Spotting of this type is generally carried out with any available capillary tubing, either hand drawn or commercial. The capillary is filled by dipping it into the solution and emptied by touching the liquid to the layer. The spot should be kept as small as possible and care should be taken that the adsorbent layer is not marred or broken (leading to deformed spots). One can obtain a desired spread of concentrations by multiple sample spotting to the same spot, for example, two applications on one spot, four on another, and so on. The use of commercial hard layers (see Table 4.2) allows the multiple spotting to be carried out with less chance of layer deformation. Such a study will yield information on the number of components present and their relative concentrations. (Note, some minor components will not be seen at the lower concentrations.)

If a large number of samples is being spotted on one layer and the adsorbent is not pre-channeled and/or pre-numbered (see Table 4.2) it may be desirable to use a template of some type as a guide for spacing the spots. Several of these templates are commercially available.

If one wishes to know the precise amount of sample applied to the layer, it is necessary to prepare the sample solution carefully and to apply a known amount to the layer. This requirement will be dealt with in Section 4.4. Techniques for applying sample streaks as used in preparative TLC will be considered in Section 4.5.

Sometimes a sample can be volatilized and allowed to condense on a thin layer. This is common when TLC is combined with GC. As the volatile components from the GC come out of the exit port, they can be condensed directly onto the layer, which is then developed with a suitable solvent. Thus, one can apply GC and TLC to the same sample with a minimum of manipulation.

Two techniques are available for concentrating a round sample spot into a thin line before chromatography. Such thin lines separate into bands rather than spots and often give sharper resolution of multi-component samples. One technique involves placing the spotted layer in a very polar solvent system such as methanol and allowing the layer to develop to a point a little above the original spot. The sample will move with the solvent and flow into a line. The layer can then be taken out, dried (remember that it takes a long time for all the methanol to be removed) and developed with the normal solvent. A second technique utilizes commercial layers (see Table 4.2), having a band of non-adsorbing support along one edge of the layer. The sample is deposited in this band. When the layer is developed, the solvent will carry the entire sample from the non-adsorbing band and deposit it as a thin line or bar at the edge of the adsorbing portion of the layer.

Solvent Choice

The choice of a solvent system has been thoroughly discussed in Chapter 3 and above in the microscope slide section. The addition of traces of acids and bases to solvent systems to reduce tailing and control pH on the layer has also been considered. The complexing agent 2,5-hexanedione has been added to solvent systems (0.5%) to facilitate the separation of inorganic ions.[7]

Normal care should be taken to carry out chromatography with pure solvents, or at least solvents that can be duplicated. Chromatographic grade solvents are available, but are expensive. They should, however, be used for precise chromatography where Rf values will be reported or when careful quantitative work is the goal. Less expensive solvents can be used on a day-to-day basis as long as one remembers that they may well contain various impurities that may change the chromatography appreciably. The reuse of a solvent mixture is not recommended, for the composition is easily changed by the selective evaporation of one of the components.

One sometimes sees, on a thin layer, the phenomenon of solvent demixing. In this case a pair (or more) of solvents that are mutually soluble in one another will separate into the two on the layer. This is common when two solvents of very different polarities are mixed. The phenomenon shows up as strange half-moon shaped spots rather than the usual round or oblong shape. Frequently, one can actually see two solvent fronts on the layer, especially when the layer is viewed against a strong light. This phenomenon has been used to enhance separations in some cases.

Development Techniques

Almost all TLC is carried out by a simple ascending technique such as shown in Figure 4.6. A number of other techniques such as circular, horizontal, and descending development are known, but appear to offer little advantage except for HPTLC; see Section 4.7. A number of topics and techniques, however, require some discussion in connection with the development process.

The development of a TLC layer is somewhat more complex than it appears. If a layer is placed in a chamber as shown in Figure 4.6, three distinct movements of solvent will occur. First, the solvent will move up through the layer, and this can be easily seen as an advancing solvent front. Second, solvent vapor will adsorb into the layer above the solvent front, thus altering its character. By the time the front arrives at the top of the layer, it will be moving into an adsorbent that is almost saturated with the more volatile component of the solvent system. The third solvent movement is the evaporation of solvent from the layer *below* the solvent front. The extent to which this takes place determines how much solvent has actually passed through the sample, since this is the sum of what one sees in the solvent front and what has evaporated below it. In most cases, one does not worry about these factors. However, for very precise work, they may be controlled with appropriate chamber saturation and layer preequilibration.

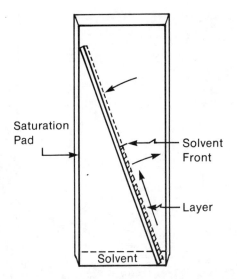

Saturation Pad

Solvent Front

Layer

Solvent

Figure 4.6 A thin layer chromatogram showing the various solvent movements. The solvent flows up through the layer, evaporates from it below the solvent front, and adsorbs into it above the solvent front.

Chromatographic Chambers and **Chamber Saturation.** TLC can be carried out in any convenient, sealable jar or container. Many designs are commercially available. Less expensive chambers can be made for 20 x 20 cm layers by smoothly cutting (or having someone cut) the top off of a standard 12 x 12 x 4 in hollow transparent glass block. The block is then closed with a glass plate.

For chromatography on large layers, one should make some attempt to saturate the chamber with solvent prior to chromatography. This will minimize the solvent evaporation as discussed above and will produce rounder and better spots. This is usually done by lining the walls of the chamber with filter paper (at least halfway around and almost to the top). The paper should be wetted with solvent, and the closed chamber should be allowed to stand a short time before the layer is placed in it.

The so-called S-chambers or sandwich chambers offer several advantages. Such a chamber is shown in Figure 4.7. In these systems, the glass plate holding the layer containing the sample is covered by a second plate. The two plates are held apart by a cardboard or Teflon spacer, which is placed around three sides of the layer. The adsorbent around the edge of the layer is removed from the area under the spacer, and the layer does not touch the cover plate. The two layers are then lightly clamped together to provide a small-volume, intact chromatographic chamber, that is dipped into solvent for development. The small volume saturates very quickly. Commercial equipment is available.

Figure 4.7 Picture of a sandwich chamber for TLC in use.

Pre-equilibration of Layers. In some cases, it is desirable to pre-equilibrate TLC layers, that is, to allow the spotted chromatogram to stand in the presence of solvent vapor for an hour or so before it is developed. This alleviates all of the problems illustrated in Figure 4.6. It is especially important for all types of LLC where the stationary liquid should be saturated with the mobile phase as well as be in equilibrium with it.

The operation can be accomplished experimentally by placing a second container, a dish or a small beaker, with a filter paper wick in the chromatographic chamber. A portion of the development solvent is placed in the container, and the chromatogram is placed in the chamber. After a period of equilibration, more of the development solvent is poured into the bottom of the chamber, and the chromatogram is allowed to develop. Solvent can be added through a hole in the chamber lid or by sliding the lid carefully to one side.

Multiple Development. The least appreciated and most important technique for improving separations is multiple development. In this technique, the chromatogram is developed once, removed from the chamber, dried, and developed again in the same solvent. It is, in fact, a means of simulating an extended distance of development, two developments of 10 cm being about equivalent to a single development of 17-18 cm. A major saving of time can be achieved; however, since the development rate de-

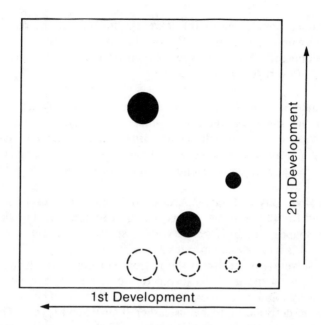

Figure 4.8 An idealized two-dimensional chromatogram with different solvents used in the two dimensions. The dashed circles represent the position of the three components after the first development and the solid spots represent their location after the second development. There would appear to be only three components in the mixture.

creases rapidly as the solvent moves up the layer, and the second development is generally more rapid than the first.

The improvement in separations is a matter of mathematics, not chromatography, and has been so treated by Thoma.[8] The equation for the prediction of the optimum number of developments for a given separation is 4.1. In general, separations are *always* improved

$$n_{opt.} = \frac{1}{\ln_e(1-Rf)} \tag{4.1}$$

by multiple development if the component spots are in the bottom half of the layer and are lessened when the spots are in the upper third of the layer (*Rf* above 0.6). This is graphically shown in Figure 4.8 where the spot locations have been calculated for two developments. Note that those spots with initial *Rf* values of 0.1 and 0.2 are spread further apart in a second development whereas those with *Rf* values of 0.6 and 0.7 are pushed together.

In light of this discussion, it is possible to define the conditions under which one can make the best possible separation with a given solvent pair.[7]

The polar component of the solvent system should be reduced to the point where the average Rf of the spots is about 0.3 after one development. The layer should then be redeveloped until the spots have an average Rf of about 0.6.

It is sometimes possible to separate certain mixtures by several developments with the same relatively non-polar solvent system that *cannot be separated* with a single development using a more polar solvent system. Much larger quantities can be separated using multiple development techniques, and consequently these are quite important in preparative TLC.

Chromatography in Shaped Areas.   It is sometimes desirable to shape the area used for TLC into the forms shown in Figure 4.9. On glass backed layers, the shapes can be formed with a spatula or a sharp instrument. Plastic backed layers can be cut out using scissors.

As the solvent emerges from the narrow portion of the layer where the sample is originally spotted, it is forced to move laterally as well as vertically (as shown by the arrows in example a of the figure). This means that the sample will be deformed into bands rather than round spots. These bands will be sharper and easier to see, and one can show a much larger number of components. Using the wedge shape (Figure 4.9d) one can see a very small amount of impurity if it runs more slowly than the major components. The technique requires a longer time than normal development.

Two–Dimensional TLC.   Two-dimensional TLC is a technique that allows the use of a large area of adsorbent for the separation of mixtures containing many components.   Furthermore, two quite different solvent

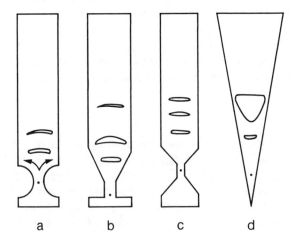

| a | b | c | d |

Figure 4.9 Chromatography in shaped areas.   Dots represent the sample origins.

systems can be used in sequence on a given mixture, thus allowing the separation of mixtures containing components with quite different polarities. The method is most useful for separating mixtures of amino acids from peptide hydrolyses and was developed for that purpose using paper chromatography.

The sample is spotted in one corner of a square layer (20 x 20 cm) and developed with one solvent system so that the mixture is resolved in a track parallel to one edge (dashed circles in Figure 4.10). The layer is removed, dried, rotated 90 deg., and placed in a second solvent system so that the spots resolved during the first development are along the bottom and are again chromatographed. The resulting components (solid spots in Figure 4.10) may be located anywhere on the layer.

One sometimes encounters compounds that decompose on a thin layer, either due to a catalytic action of the adsorbent or to the action of air on the sample. Such a possibility can be confirmed using two-dimensional TLC when both solvent systems are the same. If no decomposition has occurred, the spots will all be located on a line intersecting the original sample spot (the solid spots in Figure 4.10). If decomposition has taken place, miscellaneous spots will result off this line (from the decomposition products, open spots in Figure 4.10).

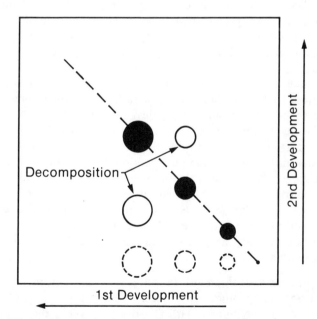

Figure 4.10 An idealized two-dimensional chromatogram using the same solvent in both directions. The dashed circles represent the location of three components after the first development and the solid spots represent their final location. The circles represent decomposition products that have formed in the course of the chromatograms.

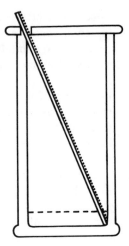

Figure 4.11 A simple arrangement for continuous thin layer chromatography.

Continuous Development. Continuous development in TLC involves the continuous flow of solvent from a reservoir (bottom of a tank, usually) through the layer with removal by some fashion at the end of the layer. In some commercial systems (Camag) the layer is horizontal. The simple apparatus depicted in Figure 4.11 represents a simple method for carrying out such a procedure. The solvent evaporates from the end of the layer into a hood.

Continuous TLC and multiple development TLC represent methods for allowing extended developments using a rather less polar solvent than one would have to use in a single, simple development. This is often desirable for the following reason. In an LSC situation, one is balancing the polarity of the solvent against the polarity of the stationary phase with the expectation that small differences in the polarities of the solute can be the basis for their separation. When the solvent is relatively polar, this can tend to swamp or override small differences between solutes and produce a poor separation. Less polar solvents will tend to enhance solute-adsorbent interactions and produce better separations. This is the reason why the best TLC results are normally achieved when the spots fall in the lower half of a layer *(Rf* less than 0.5).

Miscellaneous Development Techniques. TLC chromatograms can be developed using many other techniques. One can use a gradient elution, much as is used in column or HPLC work. In such, the polarity of the solvent is changed during the development. It is difficult to make this change in a consistent and controllable fashion, however, and the technique is less important in TLC. Circular chromatography has the characteristic of producing bands rather than spots, much like the shaped areas discussed above and is used in HPTLC as discussed at the end of this chapter. However, special equipment is needed for the development.

Rf **Values in TLC.** One of the major disadvantages of TLC is that *Rf* values are often not very reproducible, particularly when LSC is involved. Although a large number of *Rf* values have been reported in the literature, they should be regarded as valid numbers only when compared with other materials in the same study or when they are compared with a known reference substance. These difficulties result from the tremendous differences in adsorbents from the various manufacturers, as well as their difficulties in making reproducible layers.

The following precautions should be taken when measuring *Rf* values for publication:

1. Use standard commercial adsorbents, always from the same suppliers. In fact, it would be advisable to use commercially prepared layers.

2. If layers are being prepared, they should always be prepared and activated in the same way.

3. A minimum, standard time should be allowed between the removal of the layers from a dry atmosphere and development.

4. Small, known amounts of the samples should be applied in similar quantities and from solutions of similar concentration.

5. Layers should be pre-equilibrated in a carefully saturated chamber.

6. Chromatograms should be allowed to develop a standard distance and should be overrun. An overrun is carried out in the following manner. At some set height (often 10 cm) above the origin, a line is scratched across the layer. This will stop the development at this point. After the solvent reaches this point, the layer is allowed to stand in the chamber an additional 10 min or so, so that the solvent becomes uniformly distributed over the length of the chromatographic track.

7. The average of three or more experiments should be reported.

One should be careful about identifying a compound by comparing its measured *Rf* value with one reported in the literature from another laboratory. Valid identifications can be carried out only if the unknown is compared with the reference compound *on the same layer*. If the unknown is part of a complex mixture and the known is a pure sample, problems may arise because the other components of the mixture may influence the *Rf* of the compound under investigation. The technique known as spiking (see

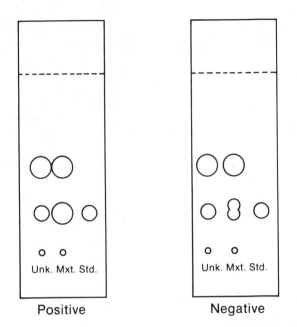

Positive   Negative

Figure 4.12 Two chromatograms showing a positive identification and a negative result when an unknown is compared with a standard and with a mixture of the standard and the unknown (the spiking technique).

Chapter 2 on GC) can be used to resolve such a problem. Thus, a small amount of the known compound should be added to the unknown mixture and the three samples should be chromatographed against one another, as shown in Figure 4.12. If the known substance is indeed in the mixture, the appropriate spot will be enlarged (positive in the figure). Otherwise the identification is negative. A positive finding at this point should be further confirmed, either by chromatography in a different system or by some visualization method (see below).

A review of standardized TLC systems for the identification of 750 drugs and poisons has been published.[9] This article includes a general discussion of reproducibility, correlation between TLC solvent system and choice of solvent mixtures, as well as an evaluation of the TLC methods for drugs and poisons.

Visualization

Visualization techniques are used to locate spots of colorless compounds on a developed layer. They may be destructive or non-destructive. Destructive methods change the sample spots irreversibly and are used for qualitative and some types of quantitative TLC. Non-destructive methods leave the sample components intact and must be used for preparative and some types of quantitative TLC. They can, of course, be used for qualitative work also. Visualization techniques may also be considered to be universal (valid for all or most organic compounds) or specific (valid only for a given type or class of compounds).

Universal Reagents or Techniques. Universal reagents are not really universal in that they can be used for all substances. Each technique has some limitation. The use of UV light on phosphor-containing layers, for example, is limited to those substances having aromatic rings or conjugated double bond systems. The iodine-chamber method discussed above works better with unsaturated or oxygen-containing compounds than with saturated hydrocarbons, and may be quite slow in some cases.

One of the major advantages cited for TLC has been that such universal reagents as concentrated sulfuric acid can be used on the completely inorganic layers (plaster of Paris bound silica gel, alumina, or kielselguhr). When the sprayed layers are heated to 100°C, organic compounds on the layers are charred to carbon and appear as black spots on a white background. The method works less well on commercial layers, which are generally bound with organic polymers. This aspect of TLC has been somewhat overstressed, however, because all black spots are alike, and subtle differences between different components do not show up. The corrosiveness of sulfuric acid can be avoided by using ammonium sulfate. Heat decomposes the ammonium sulfate to ammonia, which volatilizes, and sulfuric acid, which chars the organic compounds. Universal visualization techniques are summarized in Table 4.5. Of these methods, the UV method

Table 4.5   UNIVERSAL SPRAY REAGENTS USED IN THIN LAYER CHROMATOGRAPHY

Reagent	Composition and Use
Conc. H_2SO_4	Spray with acid and heat to 100–110° for a few minutes. Organic compounds will appear as black spots.
$(NH_4)_2SO_4$	Spray with a saturated solution of $(NH_4)_2SO_4$ in water and heat at 100–110° for a few minutes.
I_2	Spray with a 1% solution of I_2 in methanol or use I_2-chamber as described in text. Most organics will appear as brown spots when they contain an oxygen atom or purple when they do not.

is non-destructive. The iodine technique is non-destructive with about 75% of organic compounds.

Specific Techniques. For the most part, specific techniques involve spraying the layer with a reagent that will develop a color upon reacting with the sample spot. These visualization techniques were developed primarily for use in paper chromatography and, therefore, must be non-corrosive. There are two important aspects to the use of specific spray reagents in lieu of such universal methods as sulfuric acid charring. The first of these involves the functional group information that one can obtain. For example, a suspected aldehyde or ketone group may be confirmed by spraying a layer with 2,4-dinitrophenylhydrazine, or phenols may be confirmed with a ferric chloride spray, and so on. The second aspect involves the subtle shading that shows up when these spray reagents are used. No two aldehydes will give *exactly* the same color when sprayed with dinitrophenylhydrazine, and no two alkaloids will give the same color when sprayed with Dragendorff's reagent. Thus, the method can be used to confirm the presence of suspected materials in a mixture in much the same way as the spiking method mentioned above.

Some specific spray reagents are listed in Table 4.6. They have been selected primarily because of their usefulness as qualitative reagents and are destructive in nature. Many more can be found in books listed in the Bibliography.

Spray reagents should be applied in a good hood from some form of atomizer bottle, either operated by a compressed gas or a squeeze bulb. Commercial spray reagents are available in aerosol cans from many manufacturers.

Documentation of TLC Results

One important method needs to be added to those given above for documentation of the microscope size layers: the use of photography. The best method is to have a camera available that gives instant pictures in color. The photographs can be taped directly into a notebook or report and used for slide presentations. Commercial monomer solutions such as Neaton (Brinkmann) or Krylon are available that can be sprayed on a layer. After the monomer has polymerized into a film, it can be stripped off a layer and mounted in a laboratory book. Finally, a photocopier will also give a permanent record, but only in black and white with some gray tones.

When Rf values are measured as recorded data or as data for publication, the Rf of a common substance should be recorded under the exact conditions of the chromatogram. This will establish a reference system and allow one to compare data more accurately with other workers.

Table 4.6   SPECIFIC SPRAY REAGENTS FOR THIN LAYER CHROMATOGRAPHY

Reagent	Preparation and Use	Types of Compounds	Color
Aniline phthalate	Soln. A: 0.93 g aniline and 1.66 g phthalic acid in 100 mL of n-BuOH saturated with H_2O a. Spray with A b. Heat to 105° for 10 min	Reducing sugars	Various colors
Anisaldehyde in H_2SO_4 and HOAc	Soln. A: 0.5 mL of reagent in 0.5 mL conc. H_2SO_4, 9 mL of 95% EtOH, and a few drops HOAc a. Spray with A b. Heat to 105° for 25 min	Carbohydrates	Various blues
Antimony trichloride in $CHCl_3$	Soln. A: sat. soln. of reagent in alcohol-free $CHCl_3$ a. Spray with A b. Heat to 100° for 10 min c. Observe in daylight and U.V.	Steroids, steroid glycosides, aliphatic lipids, vitamin A, and others	Various colors

Table 4.6   continued

Reagent	Preparation and Use	Types of Compounds	Color
Bromcresol green	Spray with a 0.3% soln. of reagent in H_2O:MeOH (20:80) containing 8 drops of 30% NaOH per 100 mL	Carboxylic acids	Yellow spots on green
2,4-Dinitrophenylhy-drazine (2,4-DNPH)	Spray with a 0.5% soln. of reagent in 2N HCl	Aldehydes and ketones	Yellow to red spots
Dragendorff's reagent	Soln. A: 1.7 g of bismuth subnitrate in 100 mL of H_2O:HOAc (80:20) Soln. B: 40g of KI m 100 mL H_2O a. Spray with soln. made from 5 mL of A, 5 mL of B, 20 g of HOAc, and 70 mL of H_2O	Alkaloids and organic bases in general	Orange
Ferric chloride	Spray with a 1% aqueous soln. of reagent	Phenols	Various colors

Table 4.6   continued

Reagent	Preparation and Use	Types of Compounds	Color
Fluorescein: Br_2	Soln. A: 0.04% aqueous soln. of sodium fluorescein a. Spray with A b. Observe in U.V. for conjugated systems c. Expose to Br_2 d. Observe in U.V. for unsaturates	Unsaturated compounds	Yellow spots on pink
8-Hydroxyquinoline: NH_3	Soln. A: 0.5% soln. of reagent in 60% EtOH a. Expose to NH_3 b. Spray with A c. Observe in U.V.	Inorganic cations	Various colors
Ninhydrin	Soln. A: 95 mL of 0.2% reagent in BuOH plus 5 mL of 10% aqueous HOAc a. Spray with A b. Heat to 120–150° for 10–15 min	a. Amino acids b. Aminophosphatides c. Amino sugars	Blue

4.4 PREPARATIVE METHODS

The preparative separation of a mixture of similar substances by any type of chromatography is difficult and sometimes tedious. However, it is a fairly certain method, when properly applied, whereas other methods are often equally tedious and much less certain to produce the desired results. All of the chromatographic methods mentioned in this book have been used preparatively. The classic preparative method, column chromatography (Chapter 5) is still used, but is being upgraded and modified to more closely approximate HPLC (Chapter 6). GC has been used for the preparative separation of both large and small samples, but the instrumentation and columns required for large samples are complex and expensive. Preparative TLC is ideal for the separation of small samples (50 mg up to 1 g) of fairly nonvolatile compounds.

TLC methods can actually be applied to preparative problems in two ways. The first of these is preparative TLC as such; the second involves the use of TLC methods to scout conditions for column chromatography or HPLC. The former will be discussed here, and the latter will be discussed in Chapters 5 and 6.

In preparative TLC the sample to be separated is deposited in a thin line on one side of a large layer (a commercial hard layer is best) and developed in a direction perpendicular to this line so that the mixture will be resolved into bands. The bands are visualized non-destructively when they are not colored compounds, and the adsorbent containing the bands is scraped off the glass plate. The samples are then eluted from the adsorbent with a polar solvent. A typical chromatogram is shown in Figure 4.13, where a portion of each of the adsorbent bands has been scraped off. The technique is useful for separating reaction mixtures to obtain pure samples for preliminary study, for the preparation of analytical samples, in natural product work where small quantities are normal and mixtures are complex, and for the preparation of pure samples for the calibration of quantitative TLC.

Each of the steps considered previously for TLC on microscope slide layers or larger layers will be reexamined in respect to preparative work.

Commercial Layers. Thick, commercial layers of an extremely high quality are available (Table 4.5) and offer many advantages over homemade layers. They are harder; it is easier to spot the sample on them without disturbing the layer; and they are quite uniform. The newest modification is to use tapered layers with the thicker part of the layer at the top to enhance the separation. Homemade layers have one possible advantage in that they may be less expensive.

The Adsorbent. All of the normal commercial adsorbents can be, and have been, used for preparative work. As usual, silica gel has been used more than any of the others. A special series of adsorbents, the P series, is

Figure 4.13 A preparative thin layer chromatogram of a dye mixture on silica gel showing the removal of part of the adsorbent plus sample.

available through Analtech.   Two factors are important in the preparation of adsorbents for preparative layers.   First of all, the adsorbent on the layer must be clean.   If necessary, either the adsorbent or the layer can be pre-washed with methanol as described previously.   This pre-washing of the plate with an appropriate solvent can be important to remove contaminants adsorbed to the plates, such as cigarette smoke, perfume, garlic, etc.   The second is to use an adsorbent containing the 254 nm phosphor to facilitate the non-destructive visualization of the developed layers.

Layer Preparation.   The optimum thickness for preparative layers is about 1 to 1.5 mm.   Thicker layers are hard to prepare and give poorer separations.   Such layers can be prepared with most of the commercially available equipment discussed earlier, although the Camag apparatus may be a little more efficient than the Desaga apparatus.   Layers of any thickness or size can be made with the tape scheme described above.

In general, the slurries used to cast preparative layers are a little thicker than those used for thin layers. This thickening can be brought about with plaster of Paris bound layers by allowing the slurry to stand for a longer time before the layer is cast. Otherwise, more adsorbent must be added to the slurry. The layers should be allowed to dry for several hours at room temperature before activation. This will prevent cracking and case hardening. The activation is normal, that is, 100°C for at least 1 hr. It is, in fact, advisable to store layers in a non-activated state and to activate them just before use.

Sample Application. Sample application is the most critical step in preparative TLC. One must distribute fairly large volumes of sample solution (up to 2 mL) in a thin uniform band (1 to 5 mm wide) without disturbing the layer surface (excessively). The commercial hard layer was developed for this operation. Any solvent boiling between 50°C and 90°C is suitable as a sample solvent.

The sample application can be carried out in several ways. First, one can simply spot a series of spots with a capillary. This is tedious, with poor uniformity and is not recommended. With a skillful hand and a microsyringe, one can lay down a fairly thin and uniform band. A number of special pieces of apparatus have been designed for this purpose. The simplest of these is the Kontes Chromatoflex Streaker, shown in Figure 4.14. The

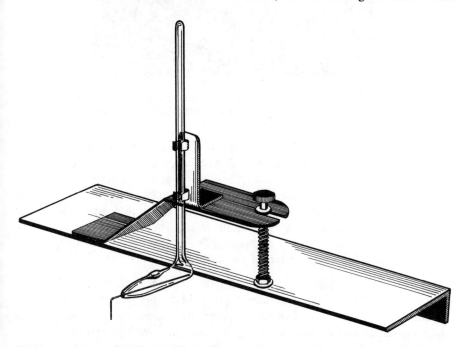

Figure 4.14 The Kontes Chromatoflex Streaker for applying a sample to a preparative TLC layer. (Reproduced through the courtesy of Kontes Glass Co.)

sample is transferred from the reservoir to the layer by a metal or glass capillary that just touches the layer as it is dragged over it.   The edge of a table is used to guide the apparatus.   The Camag Nanomat (Applied Analytical Industries) shown in Figure 4.15, which uses either a capillary pipet or a microliter syringe, is probably the most advanced device in current use. The Camag Nano-Applicator shown in Figure 4.16 is especially designed to

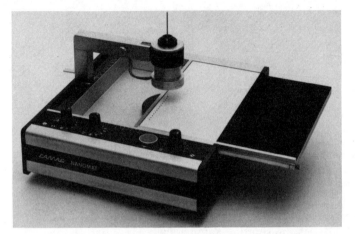

Figure 4.15 The Camag Nanomat, that applies multiple sample such that automatic scanning can be carried out.   (Reproduced through the courtesy of Applied Analytical Industries.)

Figure 4.16 The Camag Nano-Applicator for spotting very small amounts with the precision necessary for quantitation.   (Reproduced through the courtesy of Applied Analytical Industries.)

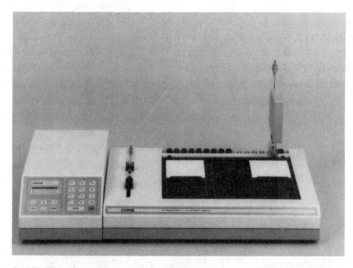

Figure 4.17 The Camag Autospotter for spotting TLC plates under micro-processor control.   (Reproduced through the courtesy of Applied Analytical Industries.)

deliver very small amounts very accurately, and the Camag Autospotter, Figure 4.17 does the spotting automatically.

Several successive applications will normally have to be made, and the layer must be dried with hot or warm gas between applications.

Sample Size. The amount of mixture that can be separated on a layer of a given size and thickness will vary widely, depending upon the type of chromatography and on the ease of separation.   About 50 mg can be separated on a 20 x 20 cm layer 1 mm thick if LSC is involved. For LLC work, 5 mg is recommended.   Quantities ranging up to 250 mg and 50 mg, respectively, have been successfully separated under ideal conditions.   For the separation of larger quantities, a number of small layers or oversize layers can be used.   Layers up to 1 cm thick and 1 m long have been used in some commercial systems to separate up to 100 g.

Development

The development of preparative layers is fairly normal, since the same solvents will yield comparable separations on thick and thin layers. However, the chamber should be well saturated with solvent vapor, and larger quantities of solvent will be needed.   The solvent system chosen should produce an average Rf of about 0.3, and the layer should be redeveloped at least once.   There is little doubt that larger samples can be separated by multi-development with a less polar system than can be separated by a single development.

Visualization

The visualization of the sample-containing bands in preparative chromatograms must be non-destructive. The best technique, by far, for use with compounds that are aromatic or have conjugated double bonds is the UV light on phosphor containing layers discussed previously. The phosphor should be the one that absorbs light at 254 nm.

If the compounds do not absorb UV light, there is no *good* method for visualizing the bands. One method involves laying a piece of transparent tape, sticky side down, on the layers so that it will be perpendicular to the sample bands. The tape will pick up some of the adsorbent containing the bands and can be removed and visualized with any spray reagent except sulfuric acid. The results are then extrapolated back to the layer. A strip along the edge of the chromatoplate can be visualized by a spray technique. This portion of the chromatogram is then discarded. One method that sometimes works on silica gel is the water spray method. The layer is completely saturated by spraying it with water. The silica gel becomes translucent, and sample bands can be seen as opaque areas in the translucent field.

Sample Recovery

The adsorbent bands containing the (hopefully) pure mixture components are then scraped off the glass plate with a spatula, a razor blade, or a rubber policeman, generally onto waxed paper or metal foil. The adsorbent is placed in a sintered glass funnel (medium or fine grade) or in a filter paper cone in a glass funnel and extracted (eluted) several times with a suitable solvent. This solvent should be just polar enough to remove the sample, perhaps one that would move the sample to an Rf of 0.8 to 0.9. If any doubt remains about the efficiency of the elution, one should wash the adsorbent with methanol or methanol-ammonium hydroxide (9:1). The two eluants should be kept separate and examined by TLC. The solvent is evaporated, and the products are isolated.

Sometimes, when the solvent is evaporated for sample recovery, a crust will form on the inside of the flask, in addition to the sample. This crust is made up of low molecular weight silicates (from silica gel) or impurities from other adsorbents. When methanol has been used to elute the samples, methyl silicates are sometimes formed and extracted. At any rate, if such a crust appears to be present, it can be dealt with in the following manner. The residue (crust and sample) should be treated with a small amount of a second solvent having a minimum polarity needed to dissolve the sample. Such a treatment will generally leave the crust behind, and the new extract can then be evaporated to recover the sample. A recrystallization may be required.

An Important Word of Caution

Almost all organic compounds decompose when allowed to remain on adsorbent layers in air and light for periods of time. Preparative TLC *must* be carried out as quickly as possible from the initial sample streaking to the final elution. If decomposition is suspected, it can be checked using the two-dimensional technique. If necessary, the chromatography can be carried out in chambers filled with nitrogen or argon and in the dark.

4.5 QUANTITATIVE TLC

Quantitative TLC requires technique and precision well above that discussed in the preceding sections and it is definitely inferior to GC when the compounds being separated are volatile or can be quantitatively converted to volatile compounds. It is also inferior to HPLC methods. Where GC or HPLC instrumentation is lacking, or where fast, less accurate results are needed, TLC can be most useful. Both non-instrumental and instrumental TLC methods are available.

Two basic techniques can be used for quantitative TLC. In the first technique the substances to be determined are assayed directly on the layer. In the second, the substances are removed from the layer and assayed, generally spectrophotometrically.

All quantitative work requires pure absorbents or layers and solvents. For this reason, among others, commercial layers are probably desirable. It may be wise to pre-wash all layers and/or adsorbents before use.

The consistent delivery of known and small amounts of solutions for sample spotting on thin layers is a difficult problem and one of the major sources of error in the method. The problem can be divided into a mechanical factor and a human factor.

The mechanical factor involves the device used to deliver the sample to the layer. Two devices are common. The first is a capillary of known inside diameter that will pick up a known amount of solvent by capillary action when it is dipped into a solution. The first examples of these were probably the Drummond microcaps, but other suppliers have similar systems (Figure 4.18). These capillaries are available to pick up volumes ranging from 1-100 μL. The filled capillary should be lowered to the layer at a 90° angle. On touching, the capillary will deliver the total contents to the layer. The capillary is held in a rubber bulb that has a hole in the end, and this hole should not be covered as the sample is picked up. It is probably not necessary to squeeze the bulb to get the entire sample on the layer.

The second mechanical device is a microsyringe calibrated in μL as described for GC. The required amount of solvent solution is expelled from the syringe to form a drop at the end of the needle. The drop is then lowered to the layer, or the layer is raised to meet the drop. When used in this manner, microsyringes are subject to an error caused by creep back. In

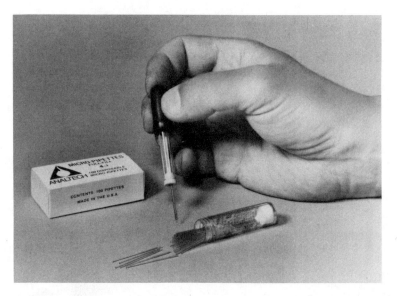

Figure 4.18 The use of pre-calibrated disposable micropipets for spotting exact amounts of sample. (Reproduced through the courtesy of Rainin Instrument Co., Inc.)

this phenomenon, the solvent creeps out of the tip of the needle and backs up the outside of the needle. Thus, a portion of the sample does not reach the layer. If this appears to be a problem, one can dip the needle in silicone oil before use.

Some of the above problems can be alleviated with the use of an apparatus designed to give reproducible sample application, such as the Camag Linomat III shown in Figure 4.19. The device uses a spray-on technique and is designed to reduce systematic errors.

The human factor in spotting can be taken care of by using standards on the same layer and by spotting the unknown and the standards in the same way. Thus, one establishes a calibration of the method and the person involved.

Assay on the Layer

When substances are assayed directly on the layer, there are no transfer or extraction errors, and the procedures are quite simple. Unfortunately, the quantization methods that one must use are not especially accurate, and overall errors vary between 5 and 10%. The quantization can be carried out in two ways: by correlating the spot size with the sample amount, and by some type of spectroscopic analysis. Instrumentation, often quite expensive, is needed for the latter.

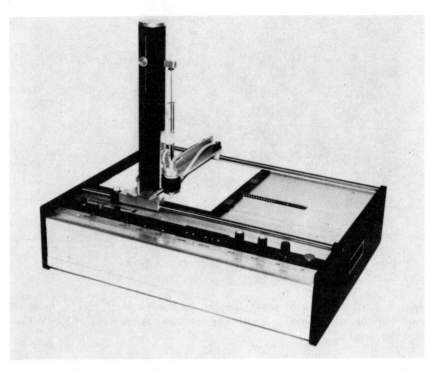

Figure 4.19 The Camag Linomat III for the spotting of samples with high precision and accuracy. (Reproduced through the courtesy of Applied Analytical Industries.)

Spot Area Measurement. There is unquestionably a relationship between the size of a spot after chromatography and the amount of solute in the spot. If standards are chromatographed on the same layer, one can often estimate the concentration of the unknown visually with about 25% accuracy.

The most precise work in this area has surely been done by Purdy and Truter.[10] These authors found that the square root of the area of a spot is directly proportional to the logarithm (to base 10) of the weight of substance present. The proportionality constant is different for different substances, and the relationship appears to be valid on quantities ranging from 1 to 80 μg on silica gel layers. A pure sample of the substance being determined must be available for calibration of the method.

The direct relationship cited above makes it possible to assay a substance without preparing a calibration curve. However, the preparation of such a curve gives some idea of the general accuracy and limits of the method. The following procedure is suggested.

1. A solution containing a known concentration of the pure substance to be assayed is prepared and diluted twice so that three solutions of known concentration are available. These concentrations should be in the general range of the concentration of the

substance in the sample to be analyzed; at best between 0.1 and 1%. Most solvents can be used to prepare these solutions. The exception is chloroform, which gives poor results.

2. The four solutions (the unknown and the three known solutions) are spotted in duplicate on a single large layer (20 x 20 cm). The spotting technique is most important and delicate. Since the spot size will be the crux of the method, it is necessary to spot the same volume of each solution in such a way that the spots are initially the same size and the layer is not disturbed (leading to spot distortions). This can best be accomplished with a microsyringe mounted in a ring stand so that the tip of the needle is just over the layer (use hard layer). The required amount of solution (5-10 μL) is forced out of the syringe, and the layer is raised to accept the drop. Multiple applications should not be made.

3. The chromatogram should be developed a premeasured distance (10-12 cm) in a well saturated chamber. The layers are dried and visualized by any appropriate technique, including charring with sulfuric acid (10 min at 120°C). An idealized chromatogram is shown in Figure 4.20.

Solvent Front								
A (mm')	113	113	232	232	113	113	36.5	36.5
$\sqrt{A}$	10.6	10.6	15.2	15.2	10.6	10.6	6.05	6.05
W (μg)	(5)	(5)	20	20	5	5	2.5	2.5
Log W	.698	.698	1.	1.	.698	.698	.398	.398

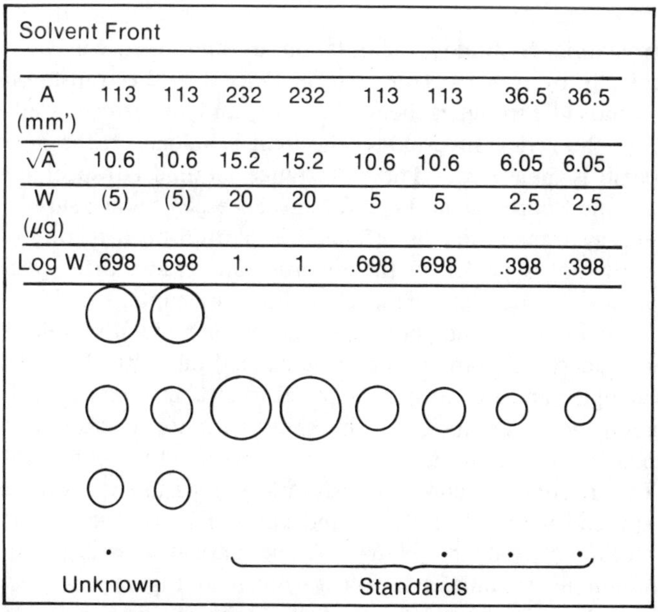

Unknown   Standards

Figure 4.20 An idealized chromatogram showing a quantitative determination by spot areas. The center spot of the unknown is being determined. All of the samples were 5 μL. The standard solutions were 0.2%, 0.1%, and 0.05% and the unknown is 0.1% also. Duplicate samples are shown, and the actual data are tabulated and superimposed on the top of the picture with units that are convenient for a determination.

4. A piece of transparent paper is then placed over the layer, and the spots are traced. The tracing is placed over a piece of millimeter graph paper and the squares in each spot are counted. A drawing instrument known as a planimeter can be used for this area measurement, but the square-counting process is less tedious than it sounds. Alternately, a photocopy can be made of the visualized chromatogram, and the spots can be cut out and weighed to determine their area.

5. The data from the three known solutions are then plotted on graph paper as the square root of the spot area vs. the logarithm of the sample weight, and a straight line is drawn through the three points. The actual position of the points with respect to the line gives an idea of the accuracy of the method and the technique. The unknown sample is then determined using the calibration line.

6. The slope of the line will remain reasonably constant for subsequent determinations, but the intercepts may change slightly from time to time, depending upon the exact conditions of the chromatography. If, with each additional analysis, a known sample is chromatographed, it can be used to determine whether a correction factor is needed.

Spectroscopic Scanning.   Spots on a thin layer can be quantized spectroscopically by transmission or reflectance. In the transmission mode, the layer is passed through a beam of light, and the transmitted energy is measured. In the reflectance mode, the light is shined on the layer and the reflected beam is measured. The reflectance method is especially effective when the sample fluoresces and the fluorescence can be measured. In either case, the energy transmitted or reflected is plotted on a recorder such that the spots show up as peaks on the recorder chart, and these peaks can be measured by any of the methods discussed in the chapter on GC.

If the compounds in the spots are colored or have ultraviolet spectra or fluorescence spectra, the analysis can be carried out with light of an appropriate wavelength, and no further chemical treatment is needed. It is rather more common, however, to spray the spots with some appropriate reagent to make them colored and carry out the analysis with visible light, perhaps filtered. This becomes a classic densitometry experiment. Frequently, the spots are sprayed with sulfuric acid and charred in an oven, and the black spots are measured by densitometry. A commercial apparatus for quantitatively scanning the visualized layers is shown in Figure 4.21. A detailed description of these methods is beyond the scope of this book.

Assay by Elution Methods.   In this second technique, one must contend with errors of transfer and extraction, but the ultimate procedures of assay are more accurate. The assay is carried out by locating the spot to be

Figure 4.21 The Camag TLC Scanner for quantitating spots on TLC and HPTLC layers. (Reproduced through the courtesy of Applied Analytical Industries.)

assayed (by a non-destructive visualization method), removing the adsorbent containing the spot from the glass plate, eluting the sample from the adsorbent, and measuring the amount of material present by ultraviolet spectrophotometry or some specific colorimetric method. The following procedure is suggested.

1. The mixture containing the substance to be determined is chromatographed on a large layer (20 x 20 cm) along with three samples of pure substance of known concentration, in much the same manner as for the spot-area method. The pure samples will establish a calibration curve.

2. The amounts of sample spotted should be known as accurately as possible and it is suggested that a microsyringe or microburette be used. However, the spotting technique itself is not especially critical.

3. The chromatogram is developed in a normal fashion and visualized non-destructively.

4. The chromatogram is marked with a sharp instrument into rectangular areas, each of that contains a spot of either the unknown or the standard. An idealized chromatogram is shown in Figure 4.22. One empty area is marked off to serve as a blank value. The rectangles should have areas as nearly identical as possible.

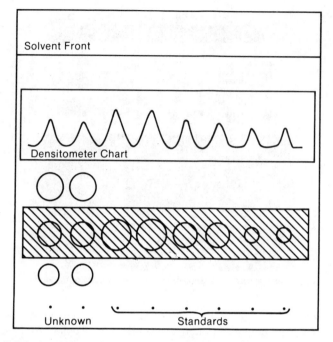

Figure 4.22 An idealized chromatogram showing a quantitative determination by densitometry. The solutions and the quantities of solutions are the same as those in Figure 4.20. The area covered by the diagonal lines was scanned by the densitometer. The areas under the peaks on the chart are directly proportional to the quantities of material present.

5. The adsorbent within each rectangle is removed from the layer as completely as possible with a razor blade or a small vacuum zone collector like that shown in Figure 4.23.

6. The sample is then eluted completely from the adsorbent with a polar solvent that will not interfere with the assay technique. The vacuum-cleaning device is designed so that it can be used for the elution step as shown. Sometimes, one can simply stir the adsorbent with solvent, centrifuge it, and analyze the supernatant liquid.

7. The eluents are then made up to a known volume and analyzed by any appropriate analytical technique. This procedure provides calibration and blank values that will not need to be done for each analysis unless the adsorbent or the chromatography is altered.

Other Methods

Radioactive samples have been measured by counting techniques, both on the layer and after elution.

Figure 4.23 A vacuum zone collector for rapid removal of adsorbent plus sample from a preparative TLC plate. (Reproduced through the courtesy of Brinkmann/Sybron Industries.)

4.6 TROUBLE SHOOTING TLC

Tailing. If the compounds being separated have long tails rather than being fairly round spots, the phenomenon is called tailing. The most common cause of tailing is too much sample or overloading. This is a natural consequence of the convex isotherms observed in most absorption processes (Figure 1.17b) and can be alleviated only by reducing sample size.

A second common cause of tailing is the lack of proper pH control on a layer. Acids and bases exist in equilibrium with their much more polar ionic carboxylate ions or ammonium ions. The pH on the layer should be such that acids are entirely in their acidic form and bases are in their amine form. Thus, acids should be chromatographed in an acidic system, and bases should be chromatographed in a basic system. This is most easily accomplished by adding a drop of acetic acid or ammonium hydroxide to the development solvent for the separation of acids or bases, respectively.

Demixing of Solvents. When two solvents having very different properties are mixed to form a solvent system, they can come unmixed on the layer, producing two solvent fronts rather than one. The two areas of the chromatogram will have quite different properties. Only solvents having fairly similar properties should be mixed.

Non-Horizontal Solvent Fronts. The solvent front is sometimes bow shaped with the low spot in the middle. This is generally caused by a lack of chamber saturation or an uneven temperature over the layer. The chamber should be carefully saturated with solvent, and care should be taken that the chamber is not placed in an air draft.

4.7 HIGH PERFORMANCE THIN LAYER CHROMATOGRAPHY

Careful study of the variables of HPLC (Chapter 6) has produced evidence that the particle size of the solid adsorbents or supports is quite critical. Although it is not yet certain just what an optimum size may be, it is quite certain that a narrow range of particle size distribution is desirable. The use of TLC adsorbents having small and narrow range particle sizes (perhaps 5-10 μm) has been termed HPTLC. It is especially good for quantitative measurements and has been used for a wide variety of samples. Figure 4.24 shows such a chromatoplate for linear development. Since the dense layers develop more slowly, they are sometimes developed in a circular or anti-circular fashion, perhaps speeded up by the centrifugal force produced by rotation. Figures 4.25 and 4.26 show the development chambers and the developed chromatograms. The small size of the development chamber permits a close control of the chromatography such that it can be treated mathematically.[11]

4.8 PAPER CHROMATOGRAPHY

Paper chromatography or PC is essentially TLC on a thin layer of cellulose or paper. The technique was invented well before TLC and has been used effectively for many years for the separation of polar biological molecules such as amino acids, sugars, and nucleotides. It is an LLC method with the liquid stationary phase, usually water, held in the fibers of the paper.

PC can best be contrasted with TLC on thin layers of powdered cellulose. PC requires no backing plates, and the paper is readily available in pure form as filter paper. Thin layers of cellulose need to be cast or purchased specifically. The fiber length is longer on paper than in the usual cellulose layers, leading to more lateral diffusion and large spots. Finally, the cellulose layers are more dense and solvent tends to flow through them more rapidly and give sharper separations.

The steps in carrying out PC are quite similar to those of partition TLC. The paper (usually Whatman No. l filter paper) is cut into strips, and the sample is spotted at one end of the strip. The chromatogram may be

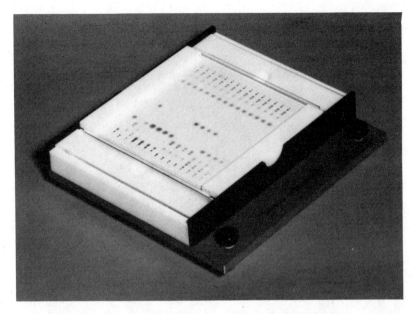

Figure 4.24 The Camag HPTLC Linear Development Chamber for TLC and HPTLC. (Reproduced through the courtesy of Applied Analytical Industries.)

Figure 4.25 The Camag Anti-Circular U-Chamber for HPTLC. (Reproduced through the courtesy of Applied Analytical Industries.)

Figure 4.26 The Camag U-Chamber for exact control of HPTLC conditions. (Reproduced through the courtesy of Applied Analytical Industries.)

developed in an ascending (Figure 4.27) or a descending fashion (Figure
4.28). For the ascending method, the paper is suspended from a hook
attached to a stopper placed in a cylinder being used as the chamber. The
solvent is in the bottom of the chamber. For the descending method, a
larger chamber is more common. It contains a trough of some type held on
a support, and the paper chromatogram is dipped into the solvent in the
trough and held in place with a glass rod. In either case, the development
takes place by capillary action. Development times in PC run from 30 min
to 12 hr, depending upon the nature of the paper and the desired development distance.

The paper sheets are removed, dried, and visualized in a similar manner
to thin layers. The spray reagents listed in Table 4.6 were actually developed for PC and are suitable reagents. Sulfuric acid cannot be used, for it
would char the cellulose as well.

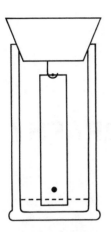

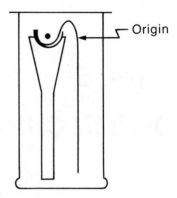
Origin

Figure 4.27 Setup for ascending paper chromatography.

Figure 4.28 Setup for descending paper chromatography.

REFERENCES:

[1]G. Guiochon, et al., *J. Chromatog. Sci.* *16* **(1978)** 152; *16* **(1978)** 470; *16* **(1978)** 598; *17* **(1979)** 368.

[2]J.J. Peifer, *Mikrochim. Acta* **(1962)** 529.

[3]This idea was first suggested by T.M. Lees and P.J. DeMuria, *J. Chromatog.* 8 **(1962)** 108 and is the basis for several commercial systems.

[4]C.B. Barrett, M.S.J. Dallas and F.B. Padley, *Chem. and Ind.* **(1962)** 1050.

[5]V. Prey, H. Berbalk and M. Kausz, *Mikrochim. Acta* **(1961)** 968.

[6]H. Brockmann, *Angew. Chem.* *59* **(1947)** 199.

[7]H. Seiler, *Helv. Chim. Acta,* *43* **(1960)** 1939; *45* **(1962)** 381.

[8]J.A. Thoma, *Anal. Chem.* *35* **(1963)** 441.

[9]A.H. Stead, R. Gill, T. Wright, J.P. Gibbs and A.C. Moffat, *Analyst, 107* **(1982)** 1106.

[10]A.J. Purdy and E.V. Truter, *Analyst, 87* **(1962)** 802; *Lab. Practice* **(1964)** 500.

[11]A. Zlatkis and R.E. Kaiser, *HPTLC-High Performance Thin Layer Chromatography,* Elsevier, New York **(1977)**.

Chapter 5

COLUMN CHROMATOGRAPHY

5.1 INTRODUCTION

Liquid chromatography carried out in large columns is the best chromatographic method for the separation of larger quantities (greater than 1 g) of mixtures. It is sometimes called **preparative liquid chromatography** or PLC. In column chromatography the mixture to be separated is introduced as a narrow band at the top of a column of adsorbent held in a glass tube, a metal tube or even a plastic tube (Chapter 1, Figures 1.9-1.12). Solvent (mobile phase) is allowed to flow through the column by gravity flow or is pushed through under pressure. The bands of solute compounds move through the column at different rates, separate, and are collected as fractions as they emerge from the bottom of the column. A graphic representation of this series of events is shown in Figure 5.1, and photos of the operation are shown in Figures 5.2-5.5. This method is an example of **elution chromatography** because the solutes are eluted from the column.

Column chromatography has been practiced as shown in Figure 5.1 for many years and is still being carried out in some laboratories. However, since about 1970, a number of the concepts and techniques developed for gas chromatography (Chapter 2) have been applied to column work. The end result of this development has been **high performance liquid chromatography,** or HPLC, probably the most sophisticated and powerful of all the known chromatographic methods. Four major changes were made in the classic column techniques. First, more finely divided and narrower range mesh adsorbents have been used to provide better equilibrium conditions in the system. Second, pressure systems, generally mechanical pumps, have been used to push the solvent through the finely divided adsorbent. This is necessary because of the small particle size, but it also makes the chromatography faster, thus minimizing diffusion. Third, detectors have been developed so that a continuous analysis is obtained of the substances as they emerge from the column. Such analysis data can be used to divide the fractions as they emerge and, when suitably treated, can provide quanti-

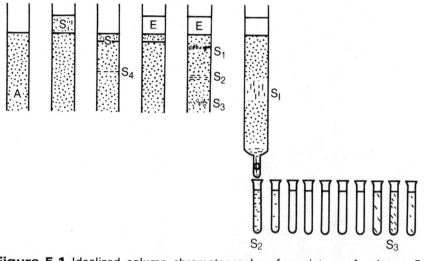

Figure 5.1 Idealized column chromatography of a mixture of solutes, S_1, S_2, S_3. The column is filled with adsorbent, A, and the sample, is added in solution. S_4 is the distance reached by the sample solvent when the sample has been completely applied, and E is the eluting solvent. As the eluting solvent leaves the column as the effluent, it is divided into fractions in the tubes shown.

tative data on the amounts of materials present. Finally, new adsorbents and new column packing techniques have been developed that provide a high degree of resolution.

While HPLC is an excellent qualitative and quantitative method for small amounts of solute, it is less satisfactory for larger quantities (although widely used as such). In most synthetic organic chemistry laboratories, various blends or hybrids of classic column techniques and HPLC have been developed for the separation of quantities ranging from 1 to 100 g. In general, such methods have a much lower resolving power than the best HPLC systems.

In this chapter, we will discuss the basic column techniques and describe some of the hybrid methods that have been developed. The chapter is planned as the second part of a continuum between the simplest of the liquid chromatographic methods, TLC (Chapter 4), and the most sophisticated, HPLC (Chapter 6). All three chapters, in turn, rest on the basic concepts discussed in Chapter 3. In the first edition of this book we discussed both liquid-solid chromatography (LSC) and liquid-liquid chromatography (LLC) as they were carried out in columns. In this edition, we will consider only LSC. Classic column LLC is little used now, and much of the work that was done using the method is done by HPLC on special bonded packings.

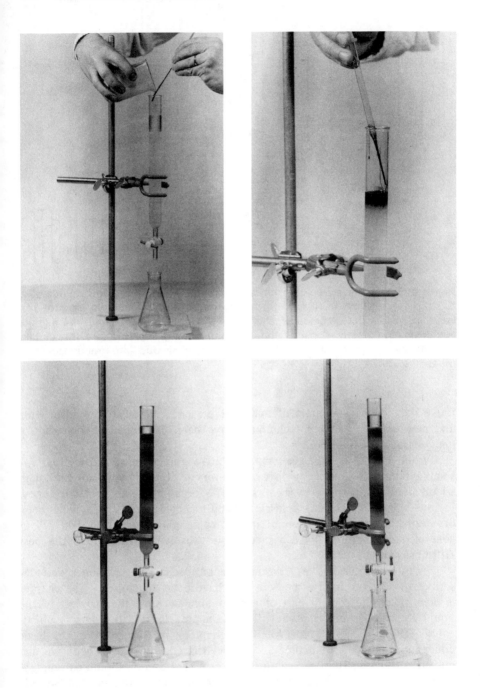

Figures 5.2-5.5 A series of pictures showing the packing a a column (5.2), the placing of a four dye sample on the column (5.3), the developed column showing the four dye bands (5.4), and the further developed column showing three bands (5.5) in the Erlenmeyer flask.

5.2 CLASSIC COLUMN LIQUID-SOLID CHROMATOGRAPHY

The nature of the adsorption process and the phenomena associated with solute separation in LSC systems were discussed in Chapter 3. In this section, we will be concerned with the materials and methods of column chromatography.

Chromatographic Columns

Chromatographic columns or tubes for gravity flow or low pressure systems are usually made of glass with some type of stopcock on the bottom to control the solvent flow. Some designs are shown in Figure 5.6. Design a is a simple buret tube. Design b comes apart at the bottom of the column for adsorbent removal and column cleaning. Design c has a glass joint at the top holding a reservoir bulb for solvent.

One of the important concepts of HPLC has been to keep the solvent volume between the adsorbent and the detector or fractionator as low as possible to prevent back mixing of the fractions after they have been separated. For this reason several commercial fittings have been designed so that a small plastic tube removes the solvent directly from the bottom of the column. Design d in Figure 5.6 has such a fitting. While most column tubes are straight, a torpedo-shaped tube such as the one in design e has also been advocated. The tube is designed so that the solvent is added and removed by small tubes held in special fittings on both ends. It is probable that such shapes improve separations by phenomena similar to those discussed for shaped area TLC in Chapter 4.

In each of the designs, there must be a support or platform of some type in the tube, above the stopcock, to hold the adsorbent in place. In designs a, b, and c, this can be a small wad of glass wool or cotton covered by a layer of clean 50-100 mesh sand about 1/8-1/4 in thick. In designs d and e and sometimes in Design b, the support may be a fritted glass disk. A layer of sand is also sometimes used on top of the frit to prevent clogging.

The overall dimensions of a column are quite variable, but the length is generally at least 10 times the internal diameter and may be as much as 100 times that value. The length-to-width ratio is largely determined by the ease or difficulty of the separation, being greater for more difficult separations. The overall size of the column and the amount of adsorbent used are determined by the weight of the solute mixture to be separated.

Amount of Mixture to Be Separated

Sometimes a crude mixture contains tars or polymeric products that will not move at all in a chromatographic system in which the major products

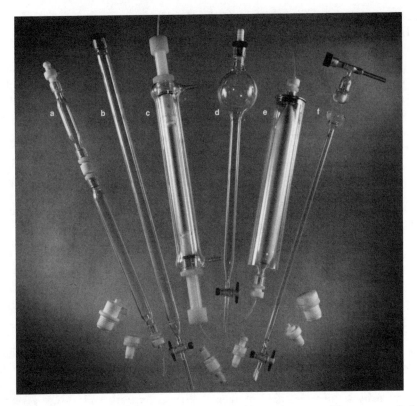

Figure 5.6 Types of glass chromatographic columns.   (Courtesy of Ace Glass, Inc.)

move. This can be seen on TLC when material, generally colored, remains at the origin and product spots move. Under these conditions, one can make a crude separation to get rid of the non-moving substances. A fairly short thick column, perhaps 10 x 2.5 cm, with a low adsorbent:solute ratio of 10:1 may be used. This is not really a separation as much as it is a purification method, much as one might use decolorizing carbon to clean a compound during recrystallization. This is sometimes done with a pre-column or guard column in HPLC.

For normal separations, a weight ratio of 30:1 has been shown to be adequate when the separation is not too difficult, as shown by widely separated spots on TLC. If the separation is more difficult, a higher ratio of adsorbent-to-solute must be used. This may be on the order of 100:1 or even 300:1, and a long thin column is more often used. A reasonable starting point would be an adsorbent-solute ratio of 100:1 in a column with an inside diameter of 2 cm when TLC shows adjacent spots.

The Adsorbent

The adsorbent, its nature, its degree or grade of activity, and its particle size are quite important in the development of a chromatographic system. An adsorbent may be prepared and treated in various ways to modify its properties and its capacity, and a number of attempts have been made to control adsorbents and to describe procedures for producing equivalent materials. The most important attempt has been the Brockmann[1] **activity grade** procedure described below. However, there have always been problems in making equivalent materials.

The particle size of column adsorbents is usually larger than that used in TLC. Column packing is usually 63-250 μm for a gravity operated column. Columns run under pressure, either from air or pumps, generally contain particles from 40-63 μm or finer. TLC adsorbents generally pass a 250 mesh screen and have particle sizes less than 63 μm. TLC adsorbents may be used in columns under pressure.

Although many different adsorbents have been used in columns, we will discuss only alumina and silica gel, two of the most useful and readily available materials.

Brockmann Activity Grade and pH. The properties of a given adsorbent will depend primarily on its pH and its activity grade. Polar surfaces such as alumina and silica gel function through oxygenated surface sites, mainly hydroxyl groups. These attract solute molecules through a complex mixture of dipole-dipole interactions and hydrogen bonding. If these sites are already occupied by water or protonic solvents such as alcohols or amines, the surface cannot function as an adsorbent, and is said to be **deactivated.** The surfaces are cleaned or activated with heat to remove the water (the main deactivating species), and the temperature and length of treatment determine the **degree of activity.** This degree of activity has been the subject of much study, and a scale, the Brockmann Index, has been devised for alumina and applied, subsequently, to other adsorbents. The grades of the scale range from I to V, where grade I has the least water (most active) and V the most water.

The degree of activity of a given alumina may be determined in the following manner:

Fill a melting point capillary tube with the adsorbent to be investigated and wet the open end of the tube with a drop of benzene. Break the closed end of the tube and immerse the wet end in a 0.5% solution of p-phenylazoaniline in benzene for a moment. Transfer the tube to a vessel containing a shallow layer of toluene and allow the solvent to migrate by capillary action to a point near the top of the small column. Remove the tube and measure the Rf

of the colored band. The approximate Rf values for the specific grades of alumina are 0.0 for grade I, 0.13 for grade II, 0.25 for grade III, 0.45 for grade IV, and 0.55 for grade V. For silica gel, the values are 0.0 for grade I and 0.65 for grade III.

Instructions for the preparation of the various activity grades of adsorbents, and for the various pH materials are given in the discussions of individual materials below.

Alumina. Alumina (Al_2O_3) is one of the most widely used adsorbents and it is available in several modifications. It possesses such sites as Al^+, Al-OH, Al^-, $Al - OH^+$ and, depending on its preparation, also Na^+ or H^+. Almost all organic compounds except saturated aliphatic hydrocarbons are adsorbed to common, basic alumina. Alumina, however, can be treated with hydrochloric acid to convert it to an acid form or with nitric acid to convert it to a neutral form. Both basic alumina, containing aluminate centers, and acid alumina, containing chloride ions, can function as ion exchangers. Basic alumina will exchange with inorganic and organic cations, and acidic alumina will exchange with inorganic and organic anions. Acidic alumina is largely used for the separation of amino acids and acid peptides, whereas neutral alumina is used for the separation of ketosteroids, glycosides, ketals, lactones, and some esters, and for the dehydration of solvents. Basic alumina has the widest range of application. Highly polar compounds are strongly adsorbed on this material, whereas non-polar compounds (except for unsaturated hydrocarbons) are weakly bound. Acetone should not be used as an eluting solvent with highly active basic aluminas, for it will be condensed to diacetone alcohol by an aldol condensation.

Alumina for adsorption is available from a number of firms and in a number of grades and qualities (see Table 5.1). Most laboratory supply companies supply aluminum oxide or alumina in a variety of grades, some of which are intended for the preparation of activated alumina by the investigator. If one is planning to simulate a TLC process, the adsorbent that is to be used in the column should be the same as that which was used to prepare the TLC layer. Thus, if aluminum oxide G was used in TLC, the equivalent grade of alumina should be obtained from the manufacturer for use in the column.

Basic alumina, grade I, can be purchased as such (Table 5.2), or it can be prepared by heating any available alumina (basic) at 380-400°C for 3 hr, with occasional stirring. Such preparations usually contain some free alkali, but this is typical of most preparations sold as basic alumina. If it is desirable to remove the alkaline substances, this product should be boiled repeatedly with distilled water until the washings are neutral, followed by washing with methanol. Activation at 200°C will again produce an activity

Table 5.1   ACTIVITY OF COMMON ADSORBENTS

Activity	% of Water		
	Silica	Alumina	Magnesium Sulfate
I	0	0	0
II	3	5	7
III	6	15	15
IV	10	25	25
V	15	38	35

Table 5.2   COMMON ADSORBENTS FOR COLUMN
CHROMATOGRAPHY

Adsorbents	Type	Supplier
Silica gel	Neutral	E. Merck J. T. Baker Bio-Rad Tridom/Fluka Woelm
Alumina	Basic, pH 10 Neutral, pH 7.5 Acidic, pH 4	Alcoa, E. Merck J. T. Baker, Bio-Rad Tridom/Fluka, Woelm
Charcoal	0.04-0.05 Particle size	Alltech Associates Tridom/Fluka
Cellulose	Anion exchange non-ionic	Bio-Rad
Polyamide	---	Bio-Rad
Polystyrene	$50-10^6$ Å Pore size	Bio-Rad, Waters

grade I material.   This material is still to be regarded as basic alumina.
Since grade I adsorbents may be too active (causing polymerizations,
dehydrations, etc.), grades with lesser activity are generally used.

Alumina of grades II, III, IV, and V can be prepared by adding 3, 6, 10,
and 15% water to the grade I adsorbent.   In practice, this is best achieved
by adding the water to a clean beaker or wide-mouthed bottle, swirling the
container to distribute the water over the walls of the vessel, and adding the
adsorbent immediately with the swirling motion being continued.   The
adsorbent should then be transferred to a powder blender or the distilling
flask of a flash evaporator and blended or rotated for at least 1 hr.

Neutral alumina can be purchased as such (see Table 5.2) or it can be
prepared[2] in the following manner.   Normal activity alumina is suspended in
water and boiled.   The supernatant liquid is made just acid to litmus with
dilute nitric acid.   The boiling process is continued and sufficient nitric acid
is added so that the boiling solution will remain acidic for 10 min after the
last addition of acid.   The adsorbent is then collected by filtration and
washed with water until the washings are neutral.   It is boiled with methanol, collected, and dried at 160-200°C under reduced pressure (10 mm) for
12-16 hr.   The resulting material is presumably activity grade I.   It can be
deactivated by the addition of water to form the various other less active
grades.

Acidic alumina can be purchased as such (see Table 5.2) or it can be
prepared in the following manner.[3]   Normal activated alumina is suspended
in three to four times its volume of 1N hydrochloric acid and stirred for
10-15 min.   The supernatant and very small particles are decanted, and the
process is repeated several times.   The alumina is then collected by filtration
on a sintered glass funnel and slowly washed with water until the washings
are only slightly acidic to litmus, followed by drying at 100°C.   The resulting material is not very active and functions primarily as an anion exchanger
in the chloride form.

Silica Gel.   Silica gel (SiO_2), or silicic acid, like alumina is a commonly
used adsorbent and might be regarded as the most versatile of all adsorbents.   Although the terms silica gel and silicic acid are used interchange-
ably, they are, in fact, modifications of the same material.[4] Silica gel can be
used with all solvents, but it shows hydrogen bonding capacity with some
solutes and solvents when water is present.   This bonding character, together with the fact that it will swell and thus slow solvent flow in the presence
of water, methanol, and ethanol, causes some limitation to its general use.

Activity grade I silica gel can ordinarily be prepared by heating at
150-160°C, with occasional stirring, for 3-4 hr.   Although highly active
grades were made for many years by heating at 300°C or higher, there is
evidence for irreversible degradation when silica gel is heated above
170°C.[5] Grade I silica gel is an anhydrous product;   grades II-V are made
by adding water to a concentration of 10, 12, 15, and 20%.

Choice of an Eluting Solvent

The choice of an adsorbent and a solvent system to make a given separation was discussed extensively in Chapter 3. The choice of an adsorbent is less important since most compounds can be separated on either alumina or silica gel, but the choice of an eluting solvent is quite an important matter. A column chromatogram represents an appreciable investment in time and materials, and it is necessary to ascertain, before starting, which solvent or mixture of solvents will cause the desired separation. There are three logical approaches to this problem. The first approach is a literature search. The second approach is to attempt to apply TLC data to a column separation. The third approach is to use a general gradient elution ranging from solvents that will not move the solutes to those more polar solvents that will move them. Gradient elution will be discussed in a later section.

Literature Search. The usefulness of a literature search depends upon whether the compounds being separated are known or not and whether they have been subjected to chromatography. This is the case for much analytical chemistry, biochemistry, and drug chemistry, and useful information can be obtained from sources listed in the Bibliography in the back of the book. One should keep in mind the facts that adsorbents vary and that the purity of the solvents used can be quite crucial.

A synthetic organic chemist is more frequently working with unknown or unusual compounds for which precise literature is not available. In such situations, one can look for information on compounds of a similar size and having similar functional groups. Literature systems can then be used as starting points for further exploration.

TLC Relationship. It would be logical to assume that a TLC system can be directly extended to a column system. Since it is possible to carry out a large number of TLC experiments in a short time with a minimum expenditure of time and solvents, one should be able to define conditions for a column separation rather easily. This is a useful approach, but it is not as simple as it sounds, nor does it always work.

The first problem in transposing TLC data to columns is that one must have adsorbents in the two methods that are as near alike as possible. One should use materials from the same manufacturer that differ only in particle size and in the presence of a binder in the TLC adsorbent. The adsorbents should be activated in the same way and handled the same way in so far as possible.

The second problem is to choose a solvent system using TLC data that will *ensure* a column separation in a reasonable amount of time and with a reasonable amount of solvent. One might look at this problem in terms of the following three closely related questions.

1. What Rf on TLC will be most likely to produce a column separation?

2. What difference in the Rf values of two spots in TLC will be likely to produce a separation?

3. How much material can be separated in a given system?

There is no general set of answers to these questions, although the flash chromatography, as discussed later, apppears to be a fairly well understood and self-consistent system.

In theory, it should be possible to find a solvent or a mixture of solvents that will produce any desired Rf value for a given substance on TLC. For a column separation, one must use a solvent that gives a relatively low Rf value on TLC. This is because there is a reciprocal relationship between the Rf value and the retention volume (V_R) of a given solute, and it is the *differences in retention volumes* of two solutes that is essential for separation on a column. The retention volume is the volume of solvent needed to move the center of a solute band to the end of a given column. This reciprocal relationship is shown in equation 5.1, that can be derived from equations 1.3 and 1.4 by the elimination of $(1 + K/\beta)$. In equation 5.2, V sub M is the holdup volume of the column or the total amount of solvent surrounding the adsorbent in the column.

$$R_f = \frac{1}{1 + K/\beta} \tag{5.1}$$

$$V_R = V_M(1 + K/\beta) \tag{5.2}$$

$$V_R = \frac{V_M}{R_f} \tag{5.3}$$

In Table 5.2, we have calculated retention volumes for a series of solutes having Rf values ranging from 0.9 to 0.1, separated on a column having a holdup volume of 100 mL. In addition, differences in Rf values and differences in retention volumes have been calculated between solutes having *adjacent Rf* values. It is easy to see that the differences in the retention volumes between solutes having high Rf values (0.9 and 0.8, for example) is low (14 mL) and the difference for low Rf values (0.1 and 0.2) is quite high (500 mL). Thus, one has a better chance of making a separation if the Rf values of the solutes are low. Although differences in retention volumes are obviously essential for separations, it does not necessarily follow that low Rf values (on TLC), high retention volumes, and an appreciable difference between the retention volumes for two solutes will produce a desirable separation. This is because of the band broadening that always takes place as a solvent passes through a chromatographic system. This was dealt with in Chapter 1. If equation 1.5 is solved for W_b, which is the width of a band at the baseline, one can see in equation 5.5 that this band width is a function of the retention volume (V_R) and the number of theoretical plates in the system (N).

$$N = 16 \frac{V_R^2}{W_b^2} \tag{5.4}$$

$$W_b = 4 \frac{V_R}{\sqrt{N}} \tag{5.5}$$

Thus, as the retention volume goes up, the band become broader. If the bands of two solutes are too broad, they overlap and a clean separation is not achieved.

In summary, if the Rf values of solutes on a TLC layer are too high, there will be an insufficient difference in retention volumes to produce a separation. If they are too low, retention volumes will be high, and band broadening and subsequent overlap will prevent a good separation. The following, rather empirical procedure should allow the transposition of the TLC data to gravity flow columns which involve adsorption chromatography.

1. Find a two-component solvent system that will yield a separation on a thin layer. Such a system should consist of a more polar solvent and a less polar solvent.

2. Modify the solvent system by reducing the amount of the polar component until the solutes of interest have Rf values below 0.3 on a thin layer. It makes little difference whether a clean separation can be *seen* at this time or not, since it is known from step 1 that a separation can be achieved.

3. Use this modified solvent system as a slurry liquid to pack a column of adsorbent that is as nearly like the TLC adsorbent as possible.

4. Apply the sample to the column and develop the chromatogram with the modified solvent system.

5. Divide the column effluent into fractions and analyze them by TLC. The fractions may or may not overlap depending upon whether the column is overloaded.

6. Combine the fractions that have the same components and evaporate the solvent to obtain the separated solutes.

The procedure above may not work very well when one of the solvent components is more polar than ethyl acetate (see the eluotropic series in Table 3.1). The more polar solvents, mostly alcohols and acetone, can occupy the sites on the adsorbent surface irreversibly, deactivating the whole system. Thus step 3 above, the column packing, would produce a completely deactivated column that would not be analogous to the TLC layer.   This problem may be approached in two ways.   First, one might arbitrarily reduce the concentration of the polar solvent by a factor of 10 and hope for the best.   Second, one might thoroughly pre-equilibrate the TLC layer with the solvent system as described in Chapter 4 so that the layer more closely approximates the situation that will be present in the column.

Finally, one may use a gradient elution technique to find a suitable solvent system for a column.   This will be discussed later in the section Elution of the Chromatogram.

Preparation of the Column

Various designs of chromatographic tubes and means of holding the adsorbent in place were discussed in a previous section.   The adsorbent may be packed into the tube either by a wet process or a dry process.   In general, the wet process is easier and more often used for silica gel while the dry process is better for alumina.

In a layer of sand is placed in the column and the dry process, the adsorbent is poured into the tube in small portions.   Each portion is leveled and compacted slightly with a plunger.   Such plungers may be a rubber stopper or a wooden plug attached to the end of a glass rod or dowel. After the adsorbent is all in place, a piece of filter paper and another layer of sand is added, so that when the solvent is added the surface will not be disturbed.   The eluting solvent is then allowed to flow down over the adsorbent with the stopcock open until the solvent level is just above the top of the column (Figures 1.9-1.12).

In the wet process, the lower layer of sand is put in place and the tube is one-third filled with solvent.   The solvent used in the packing process may be the same as the one that will be used for the chromatography itself or it may be less polar.   It should *not* be more polar.   The adsorbent is made into a slurry with another portion of the solvent, and this slurry is poured into the solvent in the tube.   During the settling process, the tube may be tapped gently on all sides with a rubber stopper or a cork ring to obtain a uniform layer.   The slurry may be added in portions or all at once. The stopcock may be open or closed during the addition as long as the solvent level is not allowed to fall below the adsorbent level.   If the solvent used in the slurry is different from the one that will be used in the chromatography, the slurry solvent should be displaced with the eluting solvent before the sample is added.

Alternatively, a wet-packed column may be prepared by half filling a tube with solvent and introducing the dry adsorbent as a fine stream through a small funnel.   The adsorbent is allowed to settle while the tube is being tapped as above to give a more uniform packing.   If the adsorbent is all introduced without interruption, an excellent column usually results.   The excess solvent is drained from the tube to provide a continuous column of adsorbent and solvent, and a piece of filter paper and a layer of washed sand is added to hold the paper in place.

Solute Addition

In the usual situation, the solutes are dissolved in a small amount of some solvent (to make a 5% or greater solution), that is added to the top of the column and allowed to flow into the top of the adsorbent. More eluting solvent is added, and the chromatogram is developed. However, since the solvent used to dissolve the solutes will be present on the column, the nature of this solvent is crucial.

Under ideal conditions the solutes are easily soluble in the solvent, that will be used in the chromatography and can be used for solute introduction. When the solutes are not easily soluble in the eluting solvent, they may be dissolved in a less polar solvent if one can be found. A more polar solvent should *not* be used, for it will alter the chromatography on the column in an unknown way.

When the solutes are not very soluble in the eluting solvent or a less polar one, as is frequently the case with complex mixtures, they may be deposited on a sample of adsorbent, which is then introduced into the top of the column. This process may be accomplished in the following manner.

1. Dissolve the solutes in any convenient solvent, preferably one that is fairly volatile.

2. Add the solution dropwise to a small amount (5-10 times as much adsorbent as solute) of activated adsorbent powder in an evaporating dish or flask.

3. Put the disk on a steam bath to evaporate the solvent or, better, put the flask on a rotary evaporator for the same purpose.

4. When the solvent is gone, the powdered adsorbent containing the solutes can be added to the top of the column. In this case, no layer of sand or piece of filter paper should have been placed on the column after it was packed.

Elution or Development of the Chromatogram

In Chapter 1, we presented a discussion of the Van Deemter equation (equation 1.8), which shows the relation between the efficiency of a separation and the speed or velocity of the elution. In summary, an elution should be as fast as possible to minimize diffusion, as long as a good equilibrium between solvent, solute, and adsorbent is maintained.   A good equilibrium requires a small particle size adsorbent.   On the other hand, very small particle size means that solvent will not flow rapidly through the adsorbent.

For gravity feed columns using adsorbents in the 60-230 mesh size (63-250 μm), flow rates are generally on the order of 10-20 mL/cm^2 of column cross section/hr.   For adsorbent particles smaller than 200 mesh, some type of pumping or pressure system will be required.   Rates may then be increased to 2 mL or so a minute, or the limit of the pressure system.

It is important that solvents used in the development of chromatograms be absolutely dry and as pure as possible.   This may require that they be dried and/or redistilled before use.   After drying, they should be stored over molecular sieves.

An elution may be **isocratic, step gradient,** or **gradient.**   In an isocratic elution, the same solvent or mixture of solvents is used throughout the chromatography.   In a step gradient elution, the composition of the solvent is changed from one mixture to another in a series of steps, each one being slightly more polar than the one before.   In a true gradient elution, the composition of solvents is continuously changed from a less polar to a more polar medium.

Step Gradient.   A traditional step gradient system is shown in Table 5.3 and follows the eluotropic series, as shown in Table 3.1.   Changes are made from one step to another when it appears that nothing more will come off the column at that particular step.   When the reservoir at the top of the column is a bulb type, as shown in Figure 5.2, some mixing of the two solvents will occur from one step to the next so that the change will not be too abrupt.   Still another step series in Table 5.3 was proposed by Rabel.[6]   The advantage of the Rabel series is that the solvents are all transparent to UV light, thus the series can be used with a UV detector.

Gradient.   A gradient is established by using solvent reservoirs such as the ones shown in Figure 5.7.   A more polar solvent is added to a reservoir of less polar solvent (reservoir b) at the same rate as solvent flows into the column.   The gradient may be exponential as shown in design A or linear as shown in design B.   In design A, solvent is added to reservoir B to maintain the same level for as long as the chromatogram takes.   The volume of solvent in B controls the abruptness of the polarity change.   In design B, no additional solvent is added to the system and the two levels go down together.

The solvents in a gradient elution may be any of the various step solutions shown in Table 5.3 or the pure solvents in Table 3.1, although it would be wise not to use two solvents of widely differing polarities, for example, hexane and methanol.   Such solvent mixtures may demix or separate during the chromatography.   Commercial equipment for establishing solvent gradients in HPLC is available.

Table 5.3 CALCULATED RETENTION VOLUMES AND THEIR DIFFERENCES FOR COLUMN SEPARATION OF SOLUTES WITH VARIOUS Rf VALUES ON TLC[1]

Rf	V_T (mL)	$\triangle Rf$ [a]	$\triangle V_T$ (mL) [b]
0.9	111		
		0.1	14
0.8	125		
		0.1	17
0.7	142		
		0.1	24
0.6	166		
		0.1	34
0.5	200		
		0.1	50
0.4	250		
		0.1	83
0.3	333		
		0.1	167
0.2	500		
		0.1	500
0.1	1000		

[a] *Calculations made for a column with a holdup volume of 100 mL using equation 5.1.*

[b] *Calculated between two adjacent Rf values.*

Detection of the Resolved Substances

One of the persistent problems of column chromatography has been the monitoring of the solvent as it comes off the column to find out when a solute is emerging. This is, of course, no problem for colored compounds,

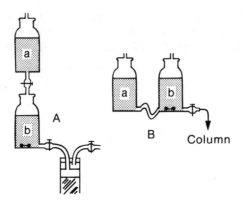

Figure 5.7 Two designs of apparatus that will provide a gradient solvent system for column development. Design A will give an exponential gradient and design B will provide a linear gradient. A stirring bar is in vessel B to provide efficient solvent mixing.

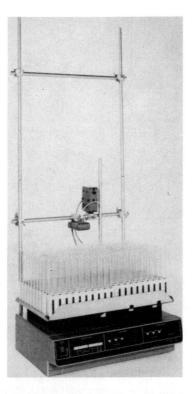

Figure 5.8 A commercial fraction collector for use in column chromatography. (Reproduced through the courtesy of Buchler Instruments.)

but most organic compounds are colorless. Traditionally, this has been done by dividing the elutent into fractions (Figure 5.1) either manually or by using a fraction collector, such as the one shown in Figure 5.8. The fractions were analyzed for solute and the concentrations were plotted against the fraction number to get a curve such as that shown in the top of Figure 5.9. The shape of the curve was then used as a guide to show which fractions should be combined for product isolation. After TLC was developed, it became common to analyze some or all of the fractions on the same layer. The bottom portion of Figure 5.9 shows an idealized chromatogram that might result from the analysis of every fifth fraction from the curve above. This is both faster and much more accurate.

The use of continuous detectors for the analysis of column eluents as developed for HPLC has revolutionized all column chromatography. The

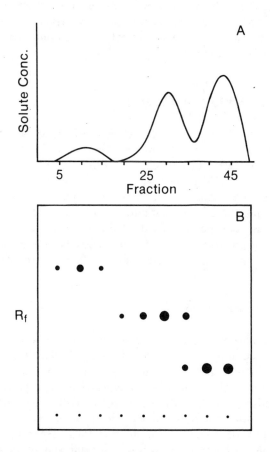

Figure 5.9 (a) An idealized curve obtained by analyzing fractions (1–45) from a column effluent. (b) The TLC picture that corresponds to the curve. Every fifth fraction was analyzed.

basis of the detection is usually UV absorption or a change in refractive index, and the results are plotted against time (which is proportional to column flow) on a recorder. The curve is similar to the one in Figure 5.9, but sharper since the analysis is continuous. The areas under the curve are proportional to the amount of solute present and may be used to obtain a quantitative analysis of a mixture. These detectors will be discussed in detail in Chapter 6 on HPLC.

Product Isolation

The column fractions that contain identical compounds (by TLC) or that appear to come from a single peak (using continuous detection) are combined, and the solvents are evaporated, preferably under vacuum. When the solvents and adsorbents are pure, the fractions should be pure. However, in many cases, one recrystallization or a crude distillation may be needed to obtain pure samples.

5.3 COLUMN - HPLC METHODS

A number of methods have been devised, primarily by organic chemists, that combine the classic methods discussed above with newer concepts and instrumentation that were developed for HPLC or GC. The goals of these methods are (1) to devise a sure way to relate Rf values from TLC with possible column separations, (2) to separate quantities ranging from 1-100 g, (3) to carry out the chromatography rapidly with minimum amounts of solvent, (4) to use cheaper standard adsorbents such as silica or alumina, and (5) to have a positive method for the analysis of the results. We will discuss three of these briefly. Details are available in the original publications.

Dry-Column Chromatography[7]

Dry-column chromatography is essentially a method for carrying out TLC in a closed column. The major advantage is that TLC results can be used to select a solvent that will ensure a separation on the column.

A "dry column" of an adsorbent having similar properties and activity grade to TLC adsorbent (but a larger particle size) is packed in a glass, or quartz, or better, in a thin Nylon tube. This process uses a vibrator to help obtain a solid, uniform column. The mixture to be separated is dissolved in a minimum amount of non-polar solvent and allowed to flow into the top layer of the dry column. Alternately, the sample may be deposited on a small amount of adsorbent (as described above) and placed on top of the column. With the bottom of the column open, the developing solvent is allowed to flow *slowly* into the top of the column. The speed is controlled by keeping the layer of solvent on top of the column small (3-5 cm high). Thus, the solute is flowing into dry, active adsorbent, a condition that closely approximates TLC. The solvent is allowed to flow until it reaches the bottom of the column. The bands of solute can be eluted from the column in a normal fashion, but, more frequently, the column is extruded from the tube and cut up to isolate solute zones. The solute is then washed from the adsorbent.

The system works best using pure solvents rather than mixtures of solvents to develop the chromatograms. Under these circumstances and, if the TLC and column adsorbents are similar, the TLC solvent can be used to develop the column. Mixed solvents are less easily used, but methods have been devised to do so.[7]

One does, of course, have the problem of finding colorless zones on the extruded column of adsorbent so that the column can be cut up for product isolation. The UV phosphor method described in Chapter 3 can be used if the solutes have ultraviolet chromophores. When a suitable phosphor is placed in the adsorbent and illuminated with UV light, the zones will show up as dark bands. Since glass does not transmit low wavelength UV light (254 nm), this technique can be applied only after the column has been extruded from a glass tube. However, with Nylon or quartz tubing, the results of the chromatogram can be visualized *during* development (Nylon and quartz transmit UV light). After the development has been completed, the Nylon tube and column can be cut apart with a sharp knife.

Flash Chromatography[8]

In flash chromatography, the developing solvent is pushed rapidly (by gas pressure) through a large diameter, but short column of wet adsorbent of a fairly closely controlled particle size. The solute zones are eluted into fractions (generally by hand rather than using a collector) and the fractions are analyzed by TLC.

Adsorbents should have particle sizes ranging from 40-60 μm (230-400 mesh). No purpose is served by smaller particles, and excessive pressures (HPLC type) would be required to push the eluting solvent through the column. The columns themselves are fitted with a needle valve system, the flow controller, such as is shown along with a column in Figure 5.10. The operating pressure can be furnished from a clean air line or a nitrogen cylinder that is connected to the controller as shown in the figure. When the system is assembled, air or nitrogen flows out of the open needle valve. When the valve is closed or partially closed, pressure is built up to produce the rapid development of the chromatogram.

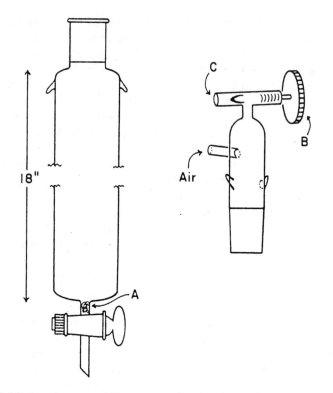

Figure 5.10 A column and flow controller for flash chromatography.

This system is probably the most self-consistent system with regard to the ability to apply TLC results to columns. A solvent system that gives an Rf of about 0.35 for the major component will normally produce a reasonable separation. The only problem arises when a very small proportion of a more polar component is used in a solvent mixture. This places the system in the range where small changes in solute composition produce large polarity changes, due to the logarithmic relationship described in Chapter 4 for TLC. Under these circumstances, the amount of polar component is reduced by one-half for column work. For example, if the ideal TLC solvent (giving an Rf of 0.35) contained 1% of ethyl acetate in hexane, one would use a column solvent containing only 0.5% of ethyl acetate.

The size of column needed depends upon the amount of sample to be separated. A summary of the column diameter, volume of eluent, typical sample loading, and common fraction size is given in Table 5.4. The actual height of the adsorbent is about 6 in in a tube 18 in long. Suppose, for example, that one wishes to separate 0.6 g of a mixture of components that,

Table 5.4   SOLUTIONS FOR STEP GRADIENTS

Traditional Series	Rabel Series[6]
Hexane	Heptane
0.1% toluene in hexane	5% chloroform in heptane
1% toluene in hexane	2% ethyl acetate in heptane
10% toluene in hexane	1% isopropyl alcohol in heptane
	5% isopropyl alcohol in heptane
Toluene	10% isopropyl alcohol in heptane
1% ether in toluene	
10% ether in toluene	
Ether	
0.1% methanol in ether	
1% methanol in ether	
10% methanol in ether	

on TLC, in the ideal solvent, produced an Rf for the major component of 0.35 and, for the nearest impurity, an Rf of 0.50. The Rf difference would then be 0.15. According to the table, one could separate the 0.6 g in a column 40 mm in diameter with 600 mL of eluent and one should collect fractions of 30 mL. Information on solutions for step gradients are given in Table 5.4 and data for the selection of the best conditions for flash chromatography are described in Table 5.5

It should be stressed that the authors describe quite specific procedures and materials. The original literature should be checked accordingly.[8]

Medium Pressure Liquid Chromatography

A number of systems have been developed using medium pressures (50-150 psi) to move the eluent through a column. In contrast, HPLC involves pressures as high as 6000 psi and higher. Generally, medium pressure systems are greatly simplified HPLC systems, frequently using non-commercial columns of conventional silica gel or alumina. A pump is usually used to deliver the solvent to the column, and the column effluent is generally collected using a fraction collector such as the one in Figure 5.4. A detector may be used to monitor the effluent, or the collected fractions may be analyzed by TLC.

Table 5.5   DATA FOR SELECTION OF CONDITIONS
FOR FLASH CHROMATOGRAPHY

Column Diameter (mm)	Vol. of Eluent[a] (mL)	Sample: Typical Loading (mg)		Typical Fraction Size (mL)
		$\Delta Rf \geqslant 0.2$	$\Delta Rf \geqslant 0.1$	
10	100	100	40	5
20	200	400	160	10
30	400	900	360	20
40	600	1600	600	30
50	1000	2500	1000	50

[a] Typical volume of eluent required for packing and elution.

Of the various systems described, the best instructions are available from Meyers and his group.[9] A schematic diagram of the system is shown in Figure 5.11. Essentially, solvent is pumped from the reservoir through the pump through a pressure and safety system through the pre-column, into the main column, and into the fraction collector. The sample is injected into the stream with a syringe through a 4-way valve. More complete descriptions of some of the components will be found in the following chapter on HPLC, and columns and packings are available.[10]

It is suggested that solvent systems that will produce an Rf of 0.2-0.3 on a comparable TLC layer will produce the desired separation. Columns were dry-packed with silica gel having a particle size of 30-60 μm or 40-60 μm. When carefully back washed after use and kept wet with hexane, the columns were usable several times. Typically, samples of 0.5-3 g could be separated on a main column 15 x 1000 mm.

Finally, we should repeat a caution given earlier in this chapter. Transpositions of solvent systems from TLC to column systems (medium pressure or HPLC) work much better when protonic solvents are avoided.

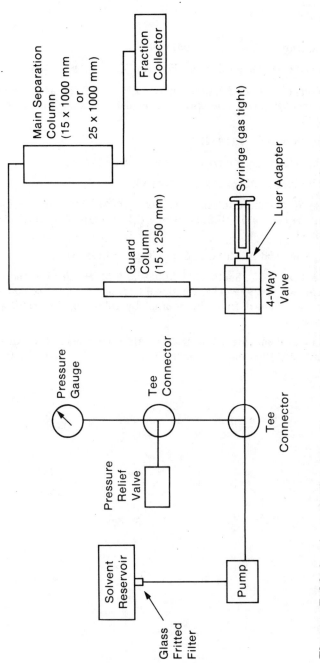

Figure 5.11 A schematic drawing of a typical medium pressure chromatography system.

REFERENCES:

[1]H. Brockmann, *Angew. Chem. 59* **(1947)** 199.

[2]T. Reichstein and C.W. Shoppee, *Disc. Faraday Soc. 7* **(1949)** 305.

[3]T. Wieland, *Hoppe-Seyl. Z. Physiol. Chem. 273* **(1942)** 24, as recorded in I.E. Bush, The Chromatography of Steroids, Pergamon Press, Inc., New York and London **(1961)** p. 352.

[4]J.J. Wren, *J. Chromatog. 4* **(1960)** 173.

[5]P. Rahn and M. Woodman, *American Laboratory, 2* **(1981)** 92.

[6]F.M. Rabel, *American Laboratory, 6* **(1974)** 33.

[7]B. Loev and M.M. Goodman, *Chem. and Ind.* **(1967)** 2026; and Progress in Separation and Purification, Vol III, E.S. Perry and C.J. van Oss, Eds., Interscience Publishers **(1970)** p. 73.

[8]W.C. Still, M. Kahn and A. Mitra, *J. Org. Chem. 43* **(1978)** 2923.

[9]A.I. Meyers, J. Slade, R.K. Smith, E.D. Mihelich, F.M. Herchenson and C.D. Liang, *J. Org. Chem. 44* **(1979)** 2247. Additional information on the components of the system can be obtained from the Journal by following the instructions described in each issue.

[10]Columns and adsorbents are available from Aldrich Chemical Co., Milwaukee, WI; Ace Glass Inc., Vineland, NJ; Amicon, Danvers, MA; and J.T. Baker, Phillipsburg, NJ.

Chapter 6

HIGH PERFORMANCE
LIQUID CHROMATOGRAPHY

6.1 INTRODUCTION

High performance liquid chromatography or HPLC is, at least for now, the ultimate method of liquid chromatography.[1] Thus, this chapter will complete our four-chapter sequence on liquid chromatography. During the past few years, the technology of HPLC and its applications have expanded tremendously and, although relatively expensive, HPLC is becoming a routine analytical and even preparative method in many laboratories. Commercial HPLC instruments consist of a highly sophisticated solvent mixing system capable of producing gradient mixtures of up to four different solvents, a pump capable of producing pressures up to 6000 (or 10,000) psi, columns containing stationary phases (or more properly, supports), and continuous detection systems of various types. Most frequently, the entire apparatus is presided over and controlled by a microprocessor. Available columns have incredible numbers of theoretical plates (more than 100,000 for a 100 cm column), and the chromatography is carried out under close to ideal conditions in such a manner that excellent separations can be obtained. Frequently, results can be obtained in minutes and interpreted quantitatively with a fair degree of precision. Samples can be separated preparatively. In Figure 6.1, a schematic diagram of an HPLC is shown.

In a sense, HPLC and GC are complementary to one another. GC has been instrumented and developed in such a way that high degrees of resolution and quantitative results are easily obtained. However, GC does require that the compounds being separated be volatile. HPLC has a corresponding limitation in that the samples must have solubility in a liquid. However, this is not such a serious limitation and, at the very least, should make HPLC applicable to most non-volatile and higher molecular weight compounds. Furthermore, HPLC can be used for inorganic compounds, most of that are not volatile. HPLC is usually carried out at room temperature; thus, thermally labile compounds can be handled readily.

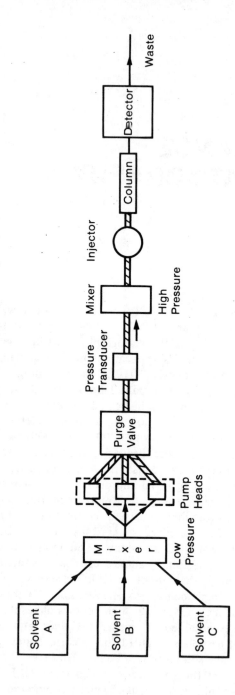

Figure 6.1 Schematic of an HPLC apparatus with three solvent capacity and with a pump with three heads. High and low pressure flows are shown.

There is one other major difference between GC and HPLC that contributes to the importance of the latter. In GC, there is little, if any, chemical interaction or association between the gas being used as a mobile phase and the sample. Looked at in another way, all gases behave in a similar manner in GC (excluding, of course, actual reactions such as oxidation). In HPLC, there is a great deal of possible interaction between the mobile phase and the solute other than a simple solubility or intermingling of molecules. Such interactions may include hydrogen bonding and ionic reactions among others. Furthermore, the properties of the mobile phase can be changed by mixing solvents in gradients of various types (Chapter 5). Thus, the number of variables in HPLC is greater, making it, at the same time, more complex and expensive, but also a more powerful separation method. One further advantage of HPLC comes from the considerably slower diffusion of solutes in liquids than in gases. An HPLC chromatogram can be stopped for a period of days and restarted without noticeable **band broadening.** This is not true for GC.

A number of technical advances have contributed to the development of HPLC (consult Figure 6.1). Pumping systems that will deliver high flow rates smoothly, that is, without the variations in pressure associated with piston action have become available, and high pressure injection systems that will allow sample injection against pressures as high as 6000 psi have been developed. One of the most remarkable developments, however, has involved the design and preparation of new forms of supports to be used in the column.

In Chapter 1, we defined two types of liquid chromatography, adsorption or liquid-solid (LSC) and partition or liquid-liquid chromatography (LLC). In general, LSC does not behave in an ideal or theoretical way (note isotherms in Figure 1.17) while LLC does. In the past, however, LLC has been carried out by depositing a stationary liquid on a support of some type and passing a liquid over the resulting column. It has always been difficult to apply the stationary phase to the support in a uniform manner and make it stay in place during the chromatography. This problem has been significantly reduced in HPLC by the preparation of bonded phase column packings. In these materials, the stationary liquid part of the LLC system is attached to the support surface by strong covalent bonds. This means that the full resolution power of the theoretically more perfect LLC system is available in a controllable form. These supports will be discussed more completely later.

The major deficiency in HPLC technology is the detector. There is no universal, highly sensitive, on-line, low-cost detector that corresponds to the flame ionization detector used in GC. A number of detector systems are available, such as the various spectroscopic detectors, fluorescence detectors, refractive index detectors, and electrochemical detectors. However, each has serious limitations.

In contrast to the TLC and classic column methods discussed in Chapters 4 and 5, HPLC is highly instrumented and can be very expensive. Many HPLC instruments, such as the one shown in Figure 6.2, are purchased as complete systems. However, it is possible to assemble an apparatus with lesser capabilities, or perhaps more suitable capabilities for a given need, from several sources at a much lower price. Some instruments were discussed in Chapter 5 under medium pressure LC systems.

The remainder of this chapter will be devoted to a discussion of the operational directions for the use of HPLC, variables to be considered in the choice of a specific solvent and column, and the technical aspects of each part of the instrumentation. The last of these sections will contain the most detail and the greatest amount of information. The chapter will be concluded with a discussion of the quantitative and preparative aspects of HPLC and a section on polymer characterization.

6.2 OPERATIONAL DIRECTIONS

All HPLC systems have the same basic parts and thus are operated in somewhat the same manner. The operation will be described as a series of steps in an **isocratic** chromatogram (one where the composition of the

Figure 6.2 An HPLC capable of ternary gradients and the associated computer system with advanced data handling and high resolution graphics. The combination gives an automated system for HPLC. (Reproduced through the courtesy of IBM Instruments, Inc.)

solvent or solvent mixture does not change). Subsequent sections will give
more information about each step. If the instrument is already turned on
and has been adjusted by the person in charge, these directions become a
series of checks.

1. The system is inspected and checked to make sure the solvent
delivery system is connected properly and that the correct column is
installed. It is important that all of the fittings are tight, that there
is enough solvent(s) in the solvent bottle(s), that the solvents are
degassed or that the **solvent degassing system** is operating to
remove air bubbles, that the **solvent filter** is installed, that the
proper detector is in place, and that the various parts are plugged
into the electrical supply.

2. The liquid flow through the column is started. This is done
either by turning on the pump after the flow controller is turned to
the proper flow rate, generally 1-2 mL/min, or by entering the
proper variables by means of a microprocessor operator station,
when the apparatus is so equipped. The **back pressure** (the pressure
required to push the mobile phase through the column) of the particular column is checked to make sure that it is not excessive (see
Table 6.1 for standard back pressures with different types of column
packing sizes and solvent mixtures). When the back pressure is too
high, the column should be changed. When handling a column, care
should be taken that **channeling** and **voids** do not occur due to
jarring.

3. HPLC is usually carried out at room temperature or at another

Table 6.1 STANDARD BACK PRESSURES WITH COLUMN TYPE

Average Particle Size (μm)	Back Pressure (psi) 25 cm Column
3	3400-5000
3	2000-3000 (15 cm)
5	2000-3000
10	900-1200
40	200-300

constant temperature. When important, the temperature of the column compartment or column block heater is checked and adjusted accordingly.

4. The mobile phase flow through the system is checked to make sure that the injection and any switching valves are adjusted properly. If the solvent flow must be measured, this can be done by collecting the flow in a volumetric flask or graduated cylinder over a specific period of time.

5. The power to the detector is turned on and adjusted to the appropriate sensitivity level. The electrical system is balanced by zeroing the output on the strip charge recorder. This adjustment is not necessary with a data integrator. The detector will probably be a UV-visible photometer type with a fixed wavelength, as this type is the simplest to use as well as the most common. When compounds that do not absorb in the UV or visible are chromatographed or when the solvent is strongly absorbing in the UV, an refractive index (RI) detector should be used (see below). Table 6.2 gives some general properties of the commonly used solvents, their UV cut-off wavelengths, and their refractive indices. After the detector is operating, the recorder trace is checked to make sure there is no excessive noise (spikes) due to bubbles caught in the pump or detector **baseline drift** due to changes in the flow rate. There should be little or no noise due to pump pulsing or electronic problems.

6. The sample in solution is filtered (normally a 0.5-2.0 μm filter) and placed in the injection valve with a microliter syringe, and the valve is turned to place the sample in the solvent stream (see Figure 6.1 and Injectors in Section 6.4). The sample could also be injected directly into the solvent stream through a special high pressure syringe or entered by a stop flow injection technique. Generally, 10 μL of solution (often as a 1 mg/mL solution in the mobile phase) is used. Each injection technique produces some baseline distortion due to a slight pressure surge. The distortion can be used to mark the injection point.

7. The peaks are recorded on a strip chart recorder, on a data integrator, or on some type of computer system with the associated printer/plotter.

8. During the chromatography, attention should be given to the gauge measuring back pressure. If the pressure increases, the system is probably clogged. If it goes down, a leak has probably developed. Most commercial HPLC systems are designed to shut down automatically when the pressure goes above a certain point.

Table 6.2   SOME PHYSICAL PROPERTIES OF SOLVENTS IN
COMMON USE IN LIQUID CHROMATOGRAPHY

Solvent	Cut-Off (nm)	Refractive Index	Dielectric Constant
n-Pentane	205	1.358	1.844
n-Heptane	197	1.388	1.924
Cyclohexane	200	1.427	2.023
Carbon tetrachloride	265	1.466	2.238
n-Butyl chloride	220	1.402	7.39
Chloroform	295	1.443	4.806
Benzene	280	1.501	2.284
Toluene	285	1.496	2.379
Dichloromethane	232	1.424	9.08
Tetrachloroethylene	280	1.938	3.42
1,2-Dichloroethane	225	1.445	10.65
2-Nitropropane	380	1.394	25.52
Nitromethane	380	1.394	35.87
n-Propyl ether	200	1.381	3.39
Ethyl acetate	260	1.370	6.02
Ether	215	1.353	4.34
Methyl acetate	260	1.362	6.68
Acetone	330	1.359	20.70
Tetrahydrofuran	225	1.408	7.58
n-Propanol	205	1.380	20.3
Ethanol	205	1.361	24.6
Methanol	205	1.329	33.6
Water	180	1.333	80.3
Acetic acid	210	1.329	6.15

INCREASING POLARITY

PTH Amino Acids

Operating Conditions

Column: IBM's Cyano 4.5 x 250 mm

Mobile Phase:

Time	NaAcetate	Acetonitrile	Methanol
0	85	15	0
5	58	30	12
7	67	15	18
15	50	25	25
20	40	30	30
22	85	15	0

Flow Rate: 1 mL/min

Detection: 254nm

Identification

1 ASP	11 TYR
2 GLU	12 VAL
3 ASN	13 PRO
4 SER	14 MET
5 THR	15 ILE
6 GLN	16 LEU
7 GLY	17 NOR-LEU
8 ALA	18 PHE
9 HYDROPRO	19 TRP
10 HIS	20 ARG

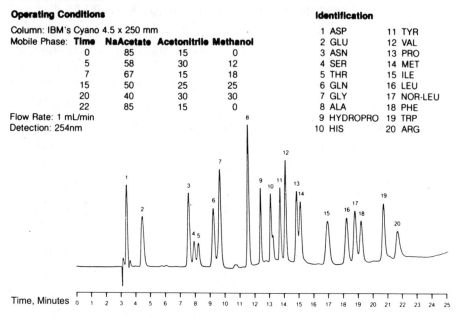

Figure 6.3 Strip chart recorder trace of an HPLC separation of amino acid derivatives. Figure 6.4 gives retention times and percentages of the peaks.

Figure 6.3 shows the chromatogram that resulted when a mixture of amino-acid derivatives was separated on a cyano bonded phase column. The output from the detector was recorded simultaneously on a strip chart recorder and a microprocessor-based data system and printer. The output from the latter is given in Figure 6.4. It is obvious that the data system gives more information in a clearer manner. From such data one can obtain retention times and **retention volumes** (as defined in Chapter 1) for compound identification as well as information from the peak heights and peak areas, that can be used to obtain quantitative results.

6.3   CHOICE OF A SYSTEM

There are three major variables to be considered in HPLC. These are, in order of increasing complexity, the detector to be used, the column packing to be chosen, and finally the mobile phase (the solvents and solvent program). These variables will be discussed in this order. The temperature is less important in HPLC as long as it is fairly constant.

```
RUN #    17

AREA%
   RT            AREA   TYPE    AR/HT      AREA%
  2.96         180570    PB     0.040      2.442
  3.51         100720    VV     0.049      1.362
  4.16         298220    VB     0.206      4.033
  7.37         418930    PV     0.203      5.666
  7.77         352470    VV     0.194      4.767
  8.07         367140    VB     0.207      4.965
  9.04         381100    BV     0.207      5.154
  9.51         545750    VB     0.191      7.381
 11.29         621540    PB     0.158      8.405
 12.02         216600    PB     0.124      2.929
 12.74         323470    BV     0.144      4.374
 13.28         297670    VV     0.128      4.026
 13.70         427590    VB     0.134      5.783
 14.35         365460    BV     0.148      4.942
 14.53         387860    VB     0.171      5.245
 16.34         415030    PB     0.322      5.613
 17.85         374880    BV     0.292      5.070
 18.50         389550    VV     0.284      5.268
 18.95         308090    VB     0.291      4.167
 20.86         355670    PB     0.259      4.810
 21.98         266150    PB     0.398      3.599


TOTAL AREA=      7394500
MUL FACTOR= 1.0000E+00
```

Figure 6.4 Microprocessor based integrator output on Figure 6.3 showing the area percentages of each of the peaks and the retention times.

The Detector

Most HPLC detectors are simply flow-through spectrophotometers set to record at some wavelength at which the solvent has little or no optical absorption and at which the samples being separated have strong absorptions. Most use ultraviolet light and measure absorption or, sometimes, fluorescence (Table 6.3).

UV detectors come in two basic types. The simplest and least expensive is a fixed wavelength detector. The wavelength is usually set at 254 nm, but can be adjusted to other wavelengths by changing filters and light sources. The variable wavelength detector can be adjusted between 190 and 700 nm so that the solvent has a minimum and the solute has a maximum absorption. These detectors generally have a wide range of sensitivities, for example, 10 or more steps from 0.001 or 0.002 to 2.0 AUFS (absorption units full scale). A number of more sophisticated UV detectors have been devised for specific purposes, but a discussion of each is beyond the scope of this book.

The most sensitive HPLC detectors are based upon fluorescence, but, of course, these can be used only with compounds that fluoresce. In order to obtain this sensitivity so that small amounts of compounds can be detected or so that valid quantitative data can be obtained, solutes are sometimes converted into fluorescing derivatives before chromatography.

Ultraviolet detectors do not work when the solvent contains UV chromophores (benzene or toluene and others) or when the solutes have no UV absorption. Under these conditions, a differential refractometer can be used, as long as the refractive index of the solvent and solutes are different. The differential refractometer measures the difference between the refractive index of a sample of pure eluting solvent and the solvent as it flows off the column, the difference being due to solutes. The refractive indices of some common solvents are given in Table 6.2. Gradient elution methods cannot be used with this type of detector since, in such systems, the solute composition and the refractive index are both changing during the chromatography. The refractive index detector is often much less sensitive than the various spectrophotometric methods.

Other detectors (Table 6.3) are the electrochemical detectors in which the redox behavior of solutes is used to measure their presence. Conductivity detectors can be used for the measurement of ionic solutes. Detectors using infrared, Raman, and mass spectrometry have been devised.

Thus, it is possible to have dual detection, with two detectors in series, to measure different properties, as long as the connections between them have a minimum volume and no detector cell is overpressured. Alternately, one can replace one detector with another, either using a valve system or by disconnecting one and connecting another.

One must use some caution in relating recorder peak sizes of various solute bands to the relative amounts of each solute present. The peak size for any given solute is a function of its spectroscopic properties and is generally not closely related to those of the other solutes. For any given solute, of course, the peak size is proportional to the concentration and can be used to make a quantitative analysis. For example, one may obtain a recorder trace from a UV detector containing two peaks of equal size and be tempted to conclude that the two solutes are present in comparable amounts. However, unless the solutes have quite similar extinction coefficients at the wavelength being used. this would not be true. One solute with a very large extinction coefficient may, in fact, be present in trace amounts.

This behavior is in contrast to some GC detectors, which respond in a similar, although not identical, manner to various solutes.

Table 6.3 TYPICAL SPECIFICATIONS FOR MOST-USED LC DETECTORS

Parameter (Units)	U.V. Visible (Absorbance)	Fluorescence	Refractive Index (RI Units)	Electrochemical (μ amp)	Conductivity (μ Mho)
Type	Selective	Selective	General	Selective	Selective
Useful with gradients	Yes	Yes	No	No	No
Upper limit of linear dynamic range	2–3	N.A.[a]	10^{-3}	2×10^{-5}	1000
Linear range (max)	10^5	$\sim 10^3$	10^4	10^6	2×10^4
Sensitivity at $\pm$ 1% noise, full scale	0.002	0.005	2×10^{-6}	2×10^{-9}	0.05
Sensitivity to favorable sample	2×10^{-10} g/mL	10^{-11} g/mL	1×10^{-7} g/mL	10^{-12} g/mL	10^{-8} g/mL
Inherent flow sensitivity[b]	No	No	No	Yes	Yes
Temperature sensitivity	Low	Low	10^{-4}°C	1.5%/°C	2%/°C

[a] N.A., not available.
[b] Because of sensitivity to temperature changes, some detectors appear to be flow sensitive.

Choice of a Column: Stationary Phase or Packing

As usual in LC, the stationary phase may be a solid surface functioning as an adsorbing medium, or a liquid surface held in place on a solid of some type. An extensive series of new stationary phase systems has been developed for HPLC, and the use of these materials has made a major contribution to the efficiency and capability of the method. Since most of these materials are based upon silica, we can center our discussion on a silica surface and its modification.

Normal and Reversed Phase Chromatography. Before we can discuss surfaces and their modification, we must again define normal and reversed phase chromatography (see Chapter 3). Of the two phases, mobile and stationary, one must always be much more polar than the other. For example, hexane as used on a silica column is much less polar than the silica surface. When the more polar of the phases is stationary, this is called **normal phase** chromatography. When the less polar phase is stationary, this is known as **reversed phase** chromatography.

Non-Bonded Column Supports. As noted in Chapter 4 and shown in Table 6.4, the silica surface is mainly a matter of hydroxyl groups attached to silicon. An unmodified silica surface, then, behaves like a stationary alcohol and interacts with the mobile phase and solute by hydrogen bonding, dipole-dipole interactions and some acid-base reactions. These multiple interactions are lumped together as **adsorption,** and chromatography on silica (and alumina, see Table 6.4) and has been extensively discussed in preceding chapters. This is normal phase adsorption chromatography or LSC.

The major difference between HPLC and the column chromatography as discussed in Chapter 5 is the particle size. In HPLC, particle sizes are in the range cf 3-10 μm and generally controlled within a very narrow range, for example, 5 ± 1 μm.

Bonded Phase Column Packings. Many new support systems developed for HPLC have been produced by *chemical modification* of the surface hydroxyl groups of silica. Some of these functional groups, as well as typical unbonded silica and alumina surfaces, are shown in a highly idealized form in Table 6.4 and Figure 6.5. In general, columns containing bonded phase packings are purchased rather than prepared in individual laboratories. The actual bonding present in the various commercial packings varies widely and is frequently a closely guarded proprietary secret. However, the bonding shown in Table 6.4 *could* exist for all. Commercial sources for a large number of packings are shown in Table 6.5.

The various coatings can be attached to the surface hydroxyl groups by a number of reactions. A representative, and one of the better reactions in

Table 6.4   STATIONARY PHASES FOR HPLC AND THEIR
CHEMICAL STRUCTURES

Unbonded Surface, Polar

Silica $\left(-Si-OH\right.$

Alumina $\left(-Al-OH\right.$

Bonded Surfaces, Non-polar

Octadecyl $\left(-Si-O-Si(CH_2)_{17}CH_3\right.$
                      $(CH_3)_2$

Octyl $\left(-Si-O-Si(CH_2)_7CH_3\right.$
                      $(CH_3)_2$

Methyl $\left(-Si-O-Si(CH_3)_3\right.$

Phenyl $\left(-Si-O-Si(CH_2)_3C_6H_5\right.$
                      $(CH_3)_2$

Amino $\left(-Si-O-Si(CH_2)_3NH_2\right.$
                      $(CH_3)_2$

Cyano $\left(-Si-O-Si(CH_2)_3CN\right.$
                      $(CH_3)_2$

Diol $\left(-Si-O-SiCH_2CHCH_2OH\right.$
                      $(CH_3)_2$ OH

use, is the reaction of a chlorosilane with hydroxyl groups on the silica. The surface of the silica contains many hydroxyl groups (on the order of 4.5 μmoles/m^2 or 27 x 10^{17} hydroxyl groups/m^2). The coated packings can be imagined as shown in Figure 6.5 and have been compared to brushes. In fact, the original packings as prepared by Halasz[2] were called

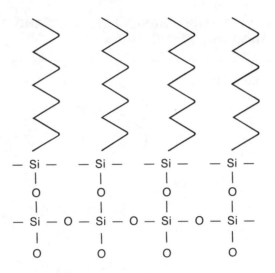

Figure 6.5 Idealized bonded phase packings.

Halasz brushes. The extent to which complete coverage of all of the hydroxyl groups is obtained will be discussed later.

In summary, either polar or non-polar coatings can be bonded to column supports, leading to either normal or reversed phase chromatography. The resulting chromatography is quite similar to LLC, but may differ in efficiency.

Choice of a Packing. The choice of the column (packing) to be used on a sample of unknown nature must be based on the general chemical nature of the solutes, their solubility properties, and their sizes. Figure 6.6 gives a selection guide. Generally, when the sample components have a molecular weight of 2000 or more and this fact is known or suspected, size exclusion chromatography (gel permeation chromatography) will be required (see Section 6.5). When the sample is polar and is soluble in an organic liquid, HPLC on a bonded packing can be used. When the sample is insoluble in organic solvents and soluble in water to give a non-neutral solution, either an ion exchange column or ion pair chromatography on a reversed phase column can be used. When the sample is soluble in water to give a neutral solution, any of the three column types can be used. The polarity of an organic soluble mixture can be ascertained by testing the solubility in less polar or more polar solvents (see Table 6.2).

A non-polar sample should be separated on a silica column. This is an important HPLC method, as it was in TLC and medium pressure column chromatography. In HPLC, adsorption is used either for separations of mixtures of isomeric compounds or for separations of mixtures of compounds of multiple polarity (differing numbers of substituent groups) that

Table 6.5   COMMERCIAL SOURCES FOR BONDED-PHASE
PACKINGS[a]

Trade Name	Supplier	Particle Sizes, μm
Partisil	Whatman	5, 10
LiChrosorb	E. M. Science	5, 10
Supelcosil	Supelco	5
Bondagel	Waters	5, 10
Zorbax	Dupont	6
Vydac	Chrompack	5, 10
Hypersil	Shandon	3, 5
Ultrasphere	Beckman	3, 5
Microsil	Micromeritics	7.5
Apex	Jones Chromatography	3, 5
-----	IBM Instruments, Inc.	5

[a] *Taken in part from R. E. Majors, J. Chromatog. Sci. 18
(1980) 488-511.*

are soluble in non-polar solvents. Generally an HPLC adsorption column is easier to degrade or damage chemically, thus deactivating or contaminating it. Since silica can be dissolved in strong base, pH's greater than 8 should be avoided. When used for a scouting run, TLC will often indicate whether a component(s) will be irreversibly adsorbed on the silica (no migration from the origin). If TLC indicates that a portion of the mixture does not move and it is desired to get rid of that portion, a solution of the mixture should be passed through a very short column of adsorbent (see Chapter 5). TLC can also indicate a possible mobile phase choice (see Chapter 3). As a corollary, TLC can be used to analyze HPLC fractions.

Partition or bonded phase HPLC is preferred for separating mixtures of homologs and very non-polar compounds that are soluble in organic solvents. Table 6.5 lists trade name and particle sizes of some of the bonded stationary phases that are presently commerically available. Figure 6.7

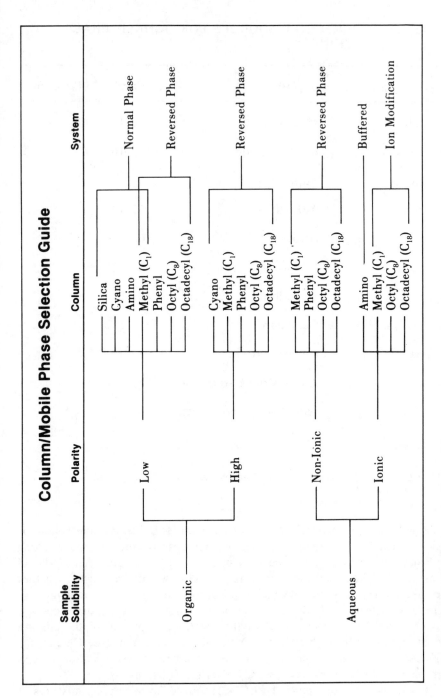

Figure 6.6 HPLC column selection guide.

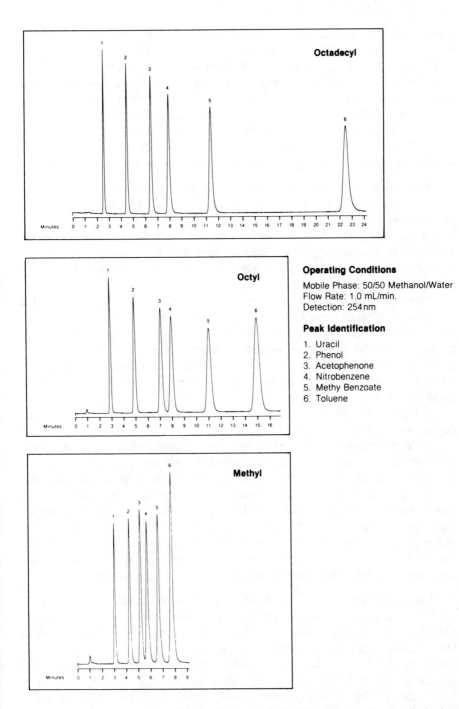

Operating Conditions

Mobile Phase: 50/50 Methanol/Water
Flow Rate: 1.0 mL/min.
Detection: 254 nm

Peak Identification

1. Uracil
2. Phenol
3. Acetophenone
4. Nitrobenzene
5. Methy Benzoate
6. Toluene

Figure 6.7 Comparison of separation capabilities of the three alkyl phase columns in the reversed phase mode.   Peak 1 is uracil, 2 is phenol, 3 is acetophenone, 4 is nitrobenzene, 5 is methyl benzoate, and 6 is toluene.

shows how the three major types of non-polar bonded packings separate a given test mixture. These are examples of reversed phase chromatography. In Figure 6.8, four of the major polar bonded packings have been used to separate a given mixture. Note that the relative position of peak 4 (benzanilide) in Figure 6.8 is shifted, depending upon the packing.

The relatively small number of different bonded packings available may suggest a limitation of HPLC on these materials. However, the nature of the packings varies so widely and the variation in the choice of a mobile phase is so infinite, that any mixture that is soluble should be separable.

Column Design. In addition to the basic chemical nature of the column packings, there are some other choices to be made. The packing may be attached to porous or hard supports that may be of varying particle sizes. In general, most HPLC is carried out on porous spheres on the order of 3-10 μm in diameter. Such materials give the best resolution with the highest capacity, but have a disadvantage in that they require a fairly high pressure for operation.

The length of an HPLC column is usually about 5-25 cm as contrasted to 0.5 m to 30 m or more for GC. This is due to the very high efficiency of HPLC (more than 100,000 theoretical plates/m as compared to 5000 plates/m for GC) as well as the higher pressures that would be required for longer columns. It is possible to couple HPLC columns in series and to recirculate the solutes through the same column. These techniques will be considered later.

Columns may be packed from commercially available packings or may be purchased already packed. The packing of gravity flow columns was discussed in Chapter 5. HPLC columns are slurry packed using high pressure (up to 6000 psi) and require considerable skill and practice for success. Since the packing materials themselves are expensive, it is probably advisable to purchase most columns to avoid the difficulties in obtaining a uniformly packed column. When the back pressure of an expensive column builds up as noted above, it is possible to open the front end of the column and replace or clean the frit. This operation may remove the clog. When not in use, columns should be stored filled with an appropriate solvent (hexane for normal phase and isopropyl alcohol for bonded phase).

Guard Columns. HPLC columns are expensive, but, unlike columns used in conventional column chromatography (Chapter 5), they can be reused many times. In order to protect the column and extend its useful life, a **guard column** or pre-column is frequently inserted between the sample inlet valve and the main column. This guard column often contains a packing similar to that used in the main column, but frequently consists of coated, larger (20-40 μm), harder spheres (pellicular) rather than the small porous materials in the main column. The pellicular guard column can be dry-packed readily in the lab. Guard columns retain solutes that may tend to deactivate or plug the main column.

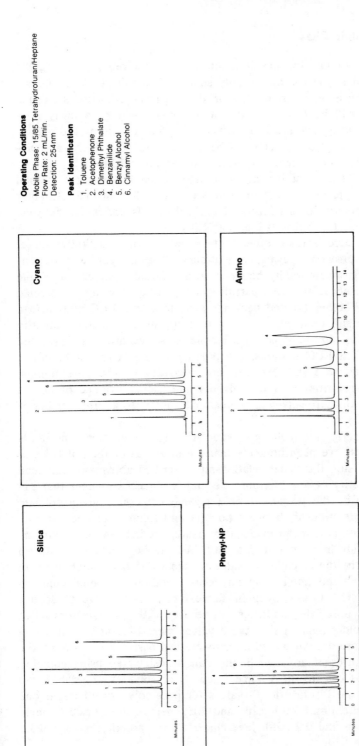

Operating Conditions

Mobile Phase: 15/85 Tetrahydrofuran/Heptane
Flow Rate: 2 mL/min.
Detection: 254nm.

Peak Identification

1. Toluene
2. Acetophenone
3. Dimethyl Phthalate
4. Benzanilide
5. Benzyl Alcohol
6. Cinnamyl Alcohol

Figure 6.8 Comparison of the separation capabilities of the four polar columns in the normal phase mode. Note the change in order of elution of the components with the cyano and amino phases.

Choice of a Mobile Phase

Once a column packing has been chosen for a given separation, the choice of a mobile phase is appreciably simplified, albeit still a challenge. When the packing is unmodified silica or alumina in an activated state, the chromatography will be LSC, and the mobile phase will be a fairly non-polar organic liquid or a mixture of such liquids. The choice of and manipulation of such adsorption systems has been discussed extensively in Chapters 3, 4, and 5. Two previously stated observations, however, can be repeated. First, TLC on active layers provides a simple and rapid method for choosing an appropriate solvent system (Chapter 5). Second (Chapter 3), in LSC the polarity of a mixture of two solvents is *not linear* between the polarities of the two solvents (see Table 3.1).

HPLC on bonded layers is quite another matter. In a qualitative sense, bonded packings provide a permanently attached "liquid layer" as a stationary phase, and chromatography between such a liquid layer and a mobile phase is more closely related to a partition or liquid-liquid system as defined in Chapter 2. Whether bonded packings provide a true LLC system in a classical sense is still debated, but, as a first approximation, the similarity can be quite useful. In the choice of a mobile phase, we are most interested in the fact that, in LLC systems, the polarity of a mixture of liquids is *linear* (contrasted to LSC) between the polarities of the pure liquids. Thus, from the polarities of pure solvents and the knowledge of how a solute behaves with these solvents, one can extrapolate to a large number of systems.

In order to carry out such an extrapolation and prediction, one needs three things. First, we need numbers that are measures of the polarities of pure solvents. Using the linear relationship described above, we can then calculate the polarity of any mixture. Second, we need a number that can be used to describe the properties of a given solute on a given column. Third, we need a relationship between polarity and column behavior.

Many attempts have been made to measure, predict, and calculate the polarities of liquids in a quantitative way, but the best values for use in HPLC appear to be the so-called P' values as measured from solubility data by Rohrschneider[3] and applied by Snyder.[1,4] Values for some common solvents used in HPLC are given in Table 6.6. These solvents do not absorb UV light appreciably and thus can be used with UV detectors set at 254 nm. It should be noted that these values were obtained from actual experiments and are not calculated on a theoretical basis. Furthermore, the polarities of mixtures of any two liquids can be calculated using equation 6.1 where ϕa and ϕb are the volume fractions of solvents a and b in the mixture and P'a and P'b are the P' values for pure solvents. Thus, if one has a mixture containing 80% hexane and 20% methylene chloride (volume fraction 0.8 hexane and 0.2 methylene chloride), the polarity, as calculated

Table 6.6   SOLVENTS FOR BONDED PHASE HPLC AND
THEIR P' VALUES[a]

Normal Phase		Reversed Phase	
Solvent	P'	Solvent	P'
*Hexane	0.1	*Water	10.2
1-Chlorobutane	1.0	DMSO	7.2
*Isopropyl ether	2.4	Ethylene glycol	6.9
Methylene chloride	3.1	*Acetonitrile	5.8
*Chloroform	4.1	*Methanol	5.1
Ethanol	4.3	Acetone	5.1
Ethyl acetate	4.4	Ethanol	4.3
*Methanol	5.1	Tetrahydrofuran	4.0
*Acetonitrile	5.8		

[a] *Taken in part from Snyder and Kirkland[1]. The *'s denote those solvents most useful for mixing systems.*

from the values in Table 6.6, will be 0.70. Actually, equation 6.1 is simply a method for averaging the polarity.

$$P' = \phi P'a + \phi P'b \qquad (6.1)$$

$$P' = 0.8 \ x \ 0.1 + 0.2 \ x \ 3.1 = 0.08 + 0.62 = 0.70$$

The expression used to describe the behavior of a given solute is the k' value or **separation factor,** defined for GC in equation 2.1. In equation 2.1, retention times were used in the calculation. For HPLC, the expression is best defined using retention volumes as in equation 6.3, where V_R is the retention volume of the solute and V_M is the volume of the mobile phase within the column.

$$k' = \frac{t_R - t_M}{t_M} \qquad (6.2)$$

$$k' = \frac{V_R - V_M}{V_M} \tag{6.3}$$

Since time and volume are directly related at a constant flow rate in both GC and HPLC, the two expressions are identical. In a sense, k' is a measure of *how much* a solute is retained by the column, and high k' values mean that the material will be held on the column a long time. The values as used in GC and HPLC and *Rf,* as used in TLC, are reciprocally related in an ideal situation, as shown in equation 6.4.

$$Rf = \frac{1}{1 + k'} \tag{6.4}$$

The volume of the mobile phase in the column, V_M, is measured experimentally by chromatographing a known substance such as uracil or phloroglucinol, which will not be retained in the column. The volume of solvent needed to move this substance through the column, or its retention volume, must then correspond to the volume of liquid in the column. This is also called the **dead volume** or zero time volume. Generally, such a substance is simply added to the mixture of solutes being chromatographed. V_M measured in this way is much more accurate than noting a baseline distortion. For example, in Figure 6.7, the compound uracil serves as a marker. In this system, uracil is not retained by the column and the volume of solvent needed to move it through the column $V_{(uracil)}$ corresponds to V_M in equation 6.3. For the solute phenol, the k' value could be calculated as follows:

$$k'_{(phenol)} = \frac{V_{(phenol)} - V_{(uracil)}}{V_{(uracil)}}$$

Since the values in Figure 6.7 are given in time, not volumes, the expression would be as in equation 6.2.

$$k'_{(phenol)} = \frac{t_{(phenol)} - t_{(uracil)}}{t_{(uracil)}} = \frac{4.5 - 2.6}{2.6} = 0.73$$

Next, we need a way to relate k' to P' so that we will be able to predict the change in k' by a given change in P'. This is given very roughly and, apparently quite empirically, by equation 6.5 where P'_1 and P'_2 are the values before and after a change, and k'_1 and k'_2 are the corresponding k' values.

$$\frac{k'_2}{k'_1} = 10^{\frac{P'_1 - P'_2}{2}} \tag{6.5}$$

Expression 6.5 applies to normal phase HPLC where more polar solvents will produce smaller k' values. For reversed phase chromatography where less polar solvents will produce smaller k' values, the expression is given in equation 6.6.

$$\frac{k'_2}{k'_1} = 10^{\frac{P'_2-P'_1}{2}}$$ (6.6)

We are now in a position to predict the k' of a solute in any solvent mixture when we know it in any given solvent with a known P' value. For example. suppose that some compound has a k' of 10 in hexane in a normal phase system. If we want to reduce the k' to 3, we can calculate the P' value for the new solvent in the following manner.

$$\frac{3}{10} = 10^{\frac{0.1-P'_2}{2}}$$

$$\log 0.3 = \frac{0.1-P'_2}{2}$$

$$0.1-P'_2 = 2 \log 0.3$$

$$P'_2 = 0.1- \log 0.3 = 1.14$$

If we then need to know what mixture of hexane and isopropyl ether will produce the P' value to give the desired k' of 3, this can be calculated using equation 6.1, where x is the volume fraction of hexane and 1-x is the volume fraction of isopropyl ether.

$$1.14 = (x) \times 0.1 + (1 - x) \times 2.4$$

$$x = 0.55$$

We would then use a mixture of 55% hexane and 45% of the ether for the separation.

Now that we can predict k' values, we must have some idea of the best value of k' to strive for in order to have the best chance for separating a solute from a mixture. This topic was considered once before in this text in relation to TLC when we needed to know the best Rf value to achieve. In general, Rf values of about 0.3 are considered best and k' values of about 5 are desirable, with a range of 2 to 10.

The overall procedure for selecting a mobile phase for bonded phase chromatography can then be summarized as follows:

1. Using Figure 6.6, choose a column packing and a type of chromatography (normal or reversed phase)

2. Using Table 6.6, choose a solvent for a first attempt. Determine the k' for this solvent (equation 2.1 or 6.2).

3. Knowing this k' and the P' for the solvent used in step 2, calculate the P' needed to produce a k' of about 5 (equation 6.5).

4. Calculate the composition of a solvent mixture that will give the desired P'.

5. Use the new solvent system. It is possible that the composition of the solvent may have to be adjusted again since the relationships (especially 6.5) are only approximate.

This procedure will provide a binary isocratic solvent system (one containing two components that will not be changed during the course of the chromatogram). The method is rather oversimplified, for certain types of solvents produce better separations of certain types of compounds. This involves solvent selectivity and is beyond the scope of this text, but is discussed in detail on pages 261-264 in Snyder and Kirkland[1] and in such recent references as Meek.[5]

The process for choosing a solvent can be approached even more empirically using Tables 6.6 and 6.7 for normal and reversed phase systems, respectively. Tables 6.6 and 6.7 are based on the fact that most HPLC systems depend upon the addition of varying amounts of solvent to a base solvent. For example, many systems used in normal phase HPLC involve the addition of amounts of more polar solvents to hexane (or heptane) as a

Table 6.7   SOLVENTS FOR NORMAL PHASE HPLC

Solvent	(Fold) Decrease in k' for 20% Addition of Solvent to Hexane
Hexane	---
Methylene chloride	1.7
Tetrahydrofuran	2.0
Ethyl acetate	2.0
Methanol	2.3
Acetonitrile	2.6

base solvent.   Table 6.6 shows the reduction in k' that will be produced by the addition of 20% of the specific solvent to hexane.   The decrease in k' is given as fold.   This simply means that k' should be divided by the number given in the table.   For example, from Table 6.6, the addition of 20% of tetrahydrofuran to hexane will reduce any k' value by one-half.   These values are also approximate.   One may use the numbers in Table 6.8 in exactly the same way for reversed phase systems.   In this case, the various solvents are added to water.   A nomogram for defining the mobile phase composition for reversed phase chromatography utilizing acetonitrile and methanol has been published,[5] and a systematic study of ternary solvent behavior using water, methanol, acetonitrile, and tetrahydrofuran has given iso-eluotropic lines for these solvents.[6,7]   The iso-eluotropic lines connect solvents of equal strength.   It is interesting that, for the column used it was shown that 1% acetonitrile compares to 0.65% ethyl ether and 0.5% isopropyl ether and 50/50 acetonitrile/water compares to 60/40 methanol/water and 37/63 tetrahydrofuran/water.[6,7]   The articles also show that ternary solvents provide a smooth transition between two limiting binary mixtures.

From the previous discussion, one might assume that an appropriate binary system could be found to make any desired separation.   This statement is not necessarily true.   The advantageous use of a ternary system for the separation of a mixture of a series of 2,4-dinitrophenylhydrazones is shown in Figure 6.9.   Figure 6.9a shows the separation with 80% methanol in water; Figure 6.9b shows the result in 75% acetonitrile in water; and Figure 6.9c shows the separation in a ternary mixture containing 20% methanol, 53% acetonitrile, and 27% water.

Gradient elution or elution with changing solvent mixtures is commonly used in HPLC.   Most commercial systems are equipped to provide binary or ternary and some even quaternary gradients.   Alternatively, one of the systems shown in Figure 5.3 can be used.   Gradient elution usually reduces the band broadening that, as shown in the Theoretical Concepts section of Chapter 1, is always present in any chromatogram.   This is sometimes called **peak focusing, solvent focusing,** or **band compression,** as illustrated in Figure 6.10.

An effective gradient elution system can be arrived at in several ways.   One might begin with a mixture containing 10% of a solvent in the base solvent (hexane or water) and increase this amount to 90% over a period of time.   This will either produce a separation or give a good idea of the polarity needed to move the solutes through the system.   One can then back up to the original 10% and approach the polarity that will move the solutes more gradually.   Finally, one can apply the methods described above for finding a binary isocratic system and approach this system with a gradient.

Figure 6.11 shows a separation that can be made using an isocratic system and Figure 6.12, a separation with a gradient system.   Figure 6.13 shows the use of a small amount of base (NH_4OH) in a system in which

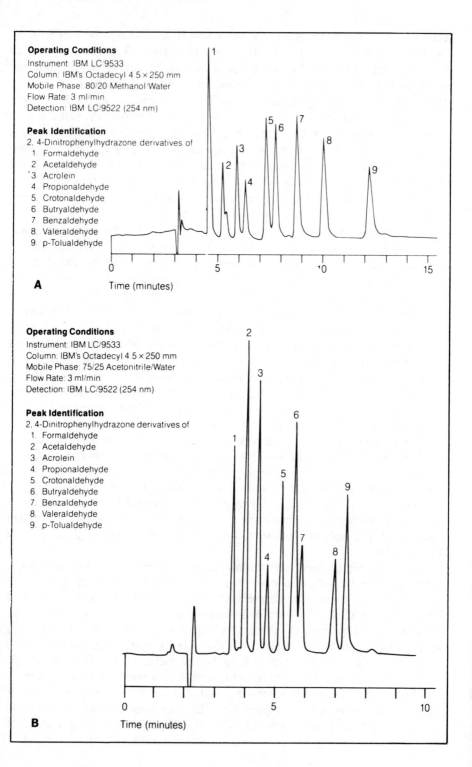

Operating Conditions
Instrument: IBM LC/9533
Column: IBM's Octadecyl 4.5 × 250 mm
Mobile Phase: 80/20 Methanol/Water
Flow Rate: 3 ml/min
Detection: IBM LC/9522 (254 nm)

Peak Identification
2, 4-Dinitrophenylhydrazone derivatives of
1. Formaldehyde
2. Acetaldehyde
3. Acrolein
4. Propionaldehyde
5. Crotonaldehyde
6. Butryaldehyde
7. Benzaldehyde
8. Valeraldehyde
9. p-Tolualdehyde

A   Time (minutes)

Operating Conditions
Instrument: IBM LC/9533
Column: IBM's Octadecyl 4.5 × 250 mm
Mobile Phase: 75/25 Acetonitrile/Water
Flow Rate: 3 ml/min
Detection: IBM LC/9522 (254 nm)

Peak Identification
2, 4-Dinitrophenylhydrazone derivatives of
1. Formaldehyde
2. Acetaldehyde
3. Acrolein
4. Propionaldehyde
5. Crotonaldehyde
6. Butryaldehyde
7. Benzaldehyde
8. Valeraldehyde
9. p-Tolualdehyde

B   Time (minutes)

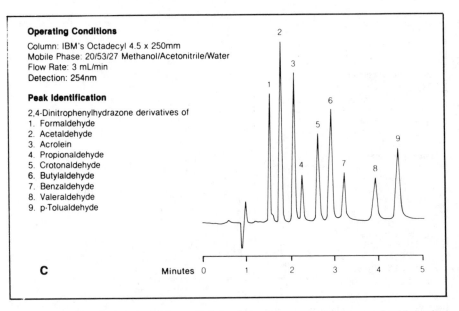

Operating Conditions

Column: IBM's Octadecyl 4.5 x 250mm
Mobile Phase: 20/53/27 Methanol/Acetonitrile/Water
Flow Rate: 3 mL/min
Detection: 254nm

Peak Identification

2,4-Dinitrophenylhydrazone derivatives of
1. Formaldehyde
2. Acetaldehyde
3. Acrolein
4. Propionaldehyde
5. Crotonaldehyde
6. Butylaldehyde
7. Benzaldehyde
8. Valeraldehyde
9. p-Tolualdehyde

C   Minutes 0   1   2   3   4   5

Figure 6.9 HPLC separation of 2,4-dinitrophenylhydrazones of aldehydes showing the incomplete resolution with (a) methanol/water or (b) acetonitrile/water and the baseline resolution with (c) a ternary mixture of the three solvents.

bases are being separated (see discussion in Chapter 3). Figure 6.14 shows the separation of very polar carbohydrates on a bonded phase column using normal solvent conditions.

Four aspects of the mobile phase liquid need to be considered in addition to its chemical composition. First, the solvents need to be chemically pure or, at least reproducible, and must be kept that way until they are used. Second, the liquids must be free of particles (particulates) that may clog the tubing or valves of the system or the column **frit** (The porous metal disk that holds in the packing). Third, the liquids must be free of dissolved gases. Fourth, the mobile phase must be mixed correctly or consistently when a mixture is used.

Prepurified solvents are commercially available for HPLC but can be expensive. Solvents may be purified by distillation or by passing them over an appropriate large column of adsorbent. They should be stored over drying agents, when appropriate, such as the various molecular sieves. When a UV detector is to be used for the HPLC, it is important to make sure that the solvents have no UV absorbing impurities.

To remove particulates, the solvents should be routinely filtered through a 5 μm or more preferable a 2 μm filter. This can be done in a typical filter flask system as shown in Figure 6.15. The filter is a special plastic disk with very small pores. Most commercial HPLC systems are

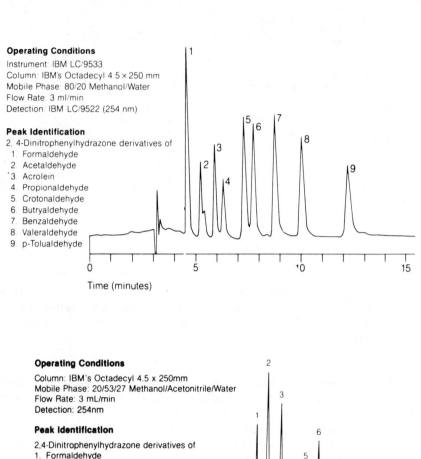

Figure 6.10 Methanol/water (a) and methanol/acetonitrile/water (c) mobile phase chromatograms from Figure 6.9 showing peak focusing from the ternary mixture.

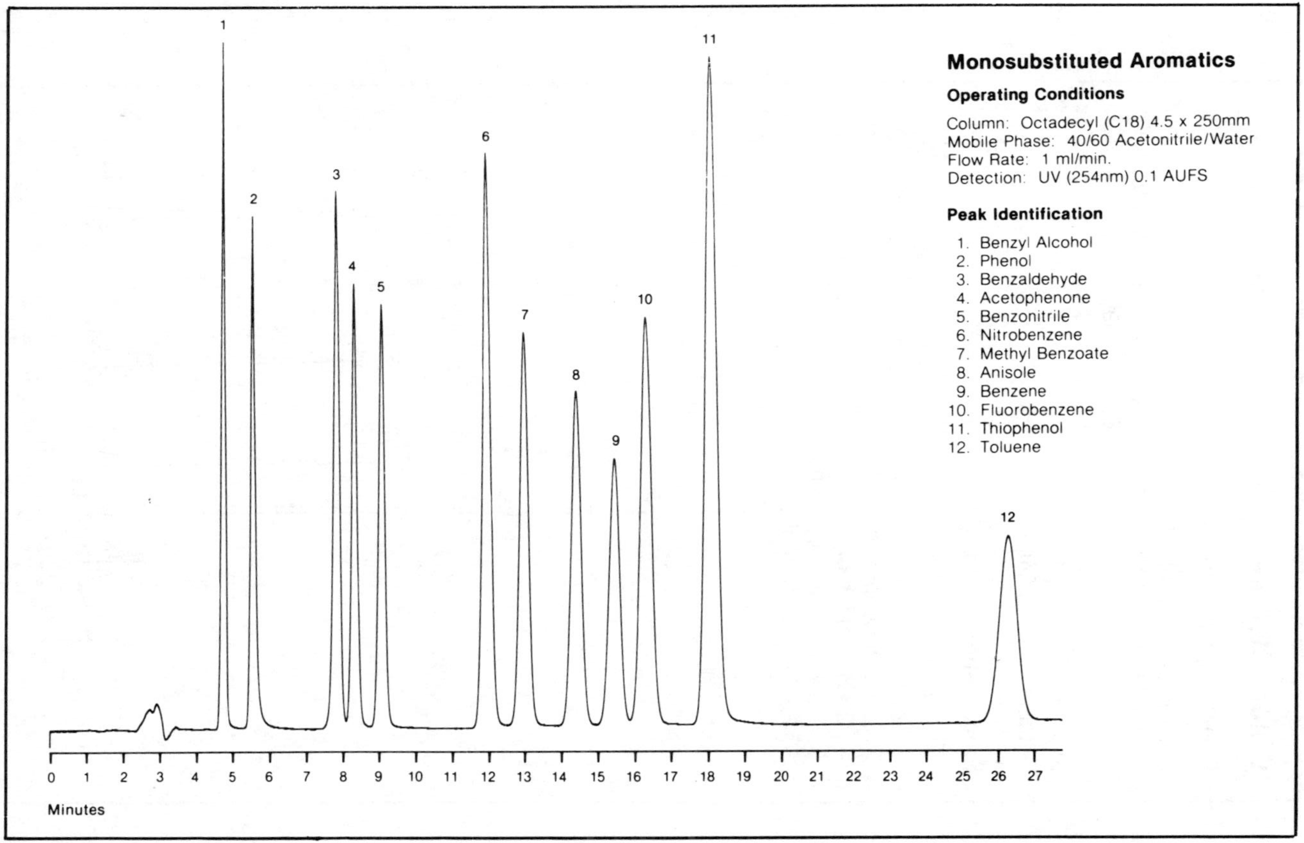
Monosubstituted Aromatics
Operating Conditions
Column: Octadecyl (C18) 4.5 x 250mm
Mobile Phase: 40/60 Acetonitrile/Water
Flow Rate: 1 ml/min.
Detection: UV (254nm) 0.1 AUFS
Peak Identification
1. Benzyl Alcohol
2. Phenol
3. Benzaldehyde
4. Acetophenone
5. Benzonitrile
6. Nitrobenzene
7. Methyl Benzoate
8. Anisole
9. Benzene
10. Fluorobenzene
11. Thiophenol
12. Toluene
0 1 2 3 4 5 6 7 8 9 10 11 12 13 14 15 16 17 18 19 20 21 22 23 24 25 26 27
Minutes

Figure 6.11 HPLC separation of mono-substituted aromatics using isocratic conditions.

Chlorophenols

Operating Conditions

Column: Octyl (C8) 4.5 x 250mm

Mobile Phase:

Time	Methanol	Water	Acetic Acid
0	39.6	59.4	1.0
10	99.0	0.0	1.0

Flow Rate: 2 ml/min.

Detection: UV (254nm) 0.2 AUFS

Peak Identification

1. Phenol
2. 2-Chlorophenol
3. 2,4-Dichlorophenol
4. 2,4,6-Trichlorophenol
5. 2,3,4,5-Tetrachlorophenol
6. Pentachlorophenol

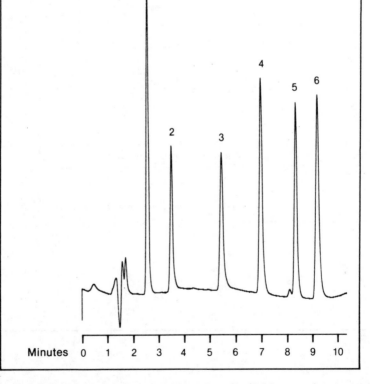

Figure 6.12 HPLC separation of chlorophenols (PCB-types) using gradient conditions.

Aromatic Amines

Operating Conditions

Column: Phenyl 4.5 x 250mm
Mobile Phase: 0.025/4.975/95
          Ammonium Hydroxide/Isopropanol/Heptane
Flow Rate: 3 ml/min
Detection: UV (254nm) 0.5 AUFS

Peak Identification

1. 2,5-Dichloroaniline
2. 2,6-Dichloro-4-nitroaniline
3. 4-Chloro-2-nitroaniline
4. 2,4-Dinitroaniline
5. 4-Nitroaniline

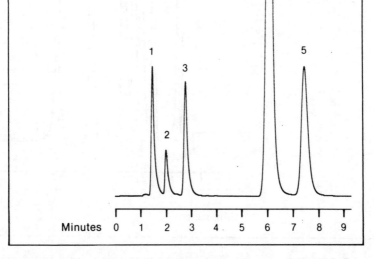

Figure 6.13 HPLC separation of aromatic amines showing the use of ammonium hydroxide to keep the amines in the basic form.

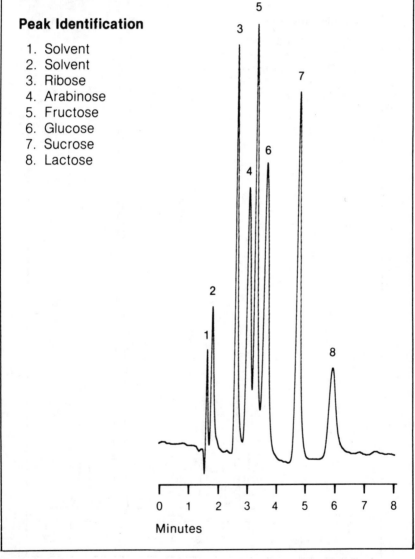

Carbohydrates

Operating Conditions

Column: Amino 4.5 x 250mm
Mobile Phase: 80/20 Acetonitrile/Water
Flow Rate: 2 ml/min.
Detection: RI 4X

Peak Identification

1. Solvent
2. Solvent
3. Ribose
4. Arabinose
5. Fructose
6. Glucose
7. Sucrose
8. Lactose

Minutes

Figure 6.14 HPLC separation of carbohydrates on a polar amino column.

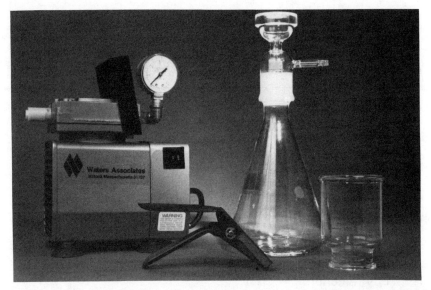

Figure 6.15 Components of a solvent filtration unit that uses either a pump or house vacuum for suction. The filter disks are either for aqueous or organic solvents. (Reproduced through the courtesy of the Millipore Corp.)

equipped with small stainless steel filtration units with a stainless steel frit, which are placed before the entrance to the pump. It is also suggested that a 2 μm filter be placed after the sample injector and before any guard or analytical column.

Dissolved gases in solvents will, unless removed, cause bubbles, that stick in the valves, pump heads, or in the detector cell. These bubbles disrupt flow and cause the detector to give false signals. The gases can be removed in several ways. Probably the easiest method is to pass helium through the solvents, rapidly for a few minutes and then slowly, during the chromatography. The helium will push out the other gases (or sparge them) and is only very slightly soluble itself. When the process is continued during the chromatography or if a blanket of helium or inert gas is placed over the solvent, it is efficiently protected against air or moisture. The difficulty with sparging is that it will selectively remove the more volatile components of a mixture and alter its composition. Thus, it should be used mainly on pure liquids. When mixtures must be degassed, they may be heated with stirring for a few minutes and then allowed to cool or they can be subjected to vacuum (15 mm) for a couple of minutes followed by blanketing with an inert gas such as argon. Most commercial instruments have built-in degassing systems.

Fresh solvent should be used insofar as it is practical, especially with solvent mixtures. Otherwise, the more volatile solvent components may evaporate and change the composition.

Table 6.8   SOLVENTS FOR REVERSED PHASE HPLC

Solvent	(Fold) Decrease in k' for 10% Addition of Solvent to Water
Water	---
Acetonitrile	2.0
Methanol	2.0
Acetone	2.1
Dioxane	2.2
Tetrahydrofuran	2.8
Isopropyl alcohol	3.0

The final aspect of mobile phase preparation applies especially to mixed solvents. For example, a 60/40 mixture of methanol/water can be correctly prepared by adding 60 mL of methanol to 40 mL of water. (Note that the final volume will be less than 100 mL.) However, adding water to 60 mL of methanol in a 100 mL volumetric flask will give a 58/42 mixture, and adding methanol to 40 mL water will give a 62/38 mixture. This error will usually change a retention time. Table 6.8 shows the large difference in k' values that can result from such small changes in the chromatography of a simple compound such as toluene. Importantly, the proportioning valves of a microprocessor controlled or modern HPLC (see below) will usually prepare the mixture correctly.

6.4 THE SYSTEM

This section will contain a more complete discussion of the various parts of an HPLC instrument and of its operation. The topics will include: (1) solvent delivery systems, (2) sample preparation and injection, (3) injection systems, (4) columns and packing, (5) detectors, (6) signal handling, and (7) fraction collectors.

Solvent Delivery Systems

The solvent delivery system consists of the solvent reservoirs, the proportioning valves, the mixing unit(s), the pump, the purge valve, and connecting tubing. Since the pump is the most important and dictates much of the rest of the system, it will be considered first.

The Pump. The pump in an HPLC system must deliver a constant and reproducible flow of solvent to the column. It must be resistant to all types of solvents, be able to reach pressures up to 6000 psi (at the present time), be essentially pulse free, and be able to deliver a metered flow of 0.01-1.0 or 0.1-20 mL/min. Furthermore, it should have a minimum **hold up volume** so that rapid solvent changeovers and efficient gradient elutions are possible. The flow rate on a pump is usually controlled by a dial on a normal pump or by a microprocessor in the more sophisticated commercial units.

Most of the pumps used in HPLC are constant flow type, either with a piston or a diaphragm, and these will be discussed below. Others such as constant pressure pumps and constant flow volume displacement pumps are beyond the scope of this book.

Constant flow pumps use either a reciprocating piston or a flexible diaphragm driven by a reciprocating piston to push the solvent through the column. Such a pump must, by its structure, lead to a pulsing of the pressure and, consequently, a pulsing of the liquid flow, much like an animal heart. Such pulsing is a major problem with this type of pump, and many methods have been developed to help alleviate or reduce it. One method is to use multiple pump heads so arranged that the pistons produce their maximum pressure at different times. Thus, double and triple heads (most efficient in reducing pump noise) are quite common, albeit expensive. These pumps still give a pulse, but a much smaller one. The pump heads are themselves quite complex in design and include **check valves** and **pump seals.**

A second method is to attach a **pulse damper** to the tube between the pump and the column. This is simply a closed space filled with a gas. At maximum pressure, some of the solvent is forced into the space, generally a long piece of capillary tubing (5 m by 0.25 mm i.d.). At minimum pressure, the solvent is returned to the system. A pulse damper, however, increases the hold-up volume within the system and causes solvent replacement and mixing to be less efficient. Finally, many commercial instruments are equipped with an electronic pulse damper controlled by a microprocessor. The microprocessor senses the pressure by means of a pressure transducer and adjusts the piston speed accordingly. These methods and combinations of them are quite successful.

Two additional problems may arise with pumps, **pump drift** and **pump noise.** Pump drift is a slow change in flow rate that is seen as a baseline drift on the recorder trace. This may be due to a slow clogging of the system or a malfunctioning pump. Pump noise or a rapidly changing baseline is due to pulsing as described above or, sometimes, gas bubbles in the system. Pumps should not be allowed to run dry, or serious abrasion or scoring may occur in the pistons or the pump heads. Mobile phases containing water and especially salt solutions should not be left in pumps, for

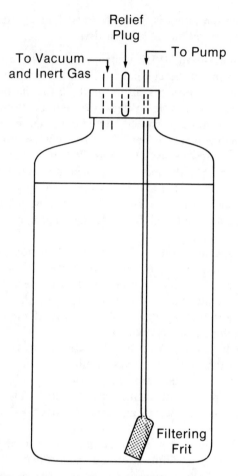

Figure 6.16 Schematic of a mobile phase reservoir that uses vacuum to remove dissolved gases and blankets with an inert gas.

corrosion or even algae growth may take place. Some of the references cited in the Bibliography in back of the book have more detailed discussions of pumps, and there is much commercial literature on the subject.

Solvent Reservoirs. In most constant flow pump systems, the solvent reservoirs are on the low pressure side of the pump so that solvent is drawn in from the reservoirs and pushed out to the column. The reservoirs may be simply filter flasks with a magnetic stirrer or they may be very complex chambers fitted with a gas purging system, a filter, and heating and temperature measuring devices, as shown in Figure 6.16. Some reservoirs are plastic coated bottles that guard against any implosion when vacuum is used in the degassing operation. A purge valve allows the solvent to flow rapidly through the tubing on the low pressure side of the pump. This is useful in making solvent changes.

Gradient Systems. The mixing of solvent to produce a binary, ternary, quaternary gradient or mixture may be done either on the low pressure side of the pump or on the high pressure side or on both sides. In its simplest form on the low pressure side, the mixing may be produced by solvent arrangements like those shown in Figure 5.3 for column chromatography. Alternately, the solvent may be drawn from a series of reservoirs through controllable proportioning valves into a small mixing chamber. These valves may be microprocessor controlled and almost any mixture or gradient, including ternary or quaternary as well as binary, can be produced.

Gradient mixing on the high pressure side is produced by more than one pump. This is usually limited to two pumps and a binary system. From the pumps, the solvent streams go to a small, usually rapidly stirred, mixing chamber and thence to the column. Since the flow rates of the pumps are completely controllable, any type of gradient may be easily produced.

Valves and Tubing. The most important aspect of all of the tubing and valve systems in HPLC is that they must have a minimum volume, the holdup volume, and no **unswept volume** (see Chapter 2). This allows facile and complete replacement of one solvent by another and efficient gradient mixing. After the solvent and solutes emerge from the column, small volume tubing to the detector lessens any back mixing of separated solutes.

Such small holdup volumes are accomplished by using small diameter tubing made of plastic, using polyethylene and sometimes Teflon, or stainless steel. The steel tubing must be used in the high pressure parts. The diameter of the tubing should not be less than 0.025 cm (0.01 in) i.d. or severe clogging may occur. Tubing should be used in as short a length as possible. Valves, as well as pumps, should have minimum holdup volumes.

Sample Preparation

Sample preparation for HPLC depends upon the source and properties of the sample. In some cases, the sample may be chemically modified to provide compounds that are easier to separate or easier to detect after separation.

In general, however, the sample is dissolved in some liquid, filtered, and injected into the solvent stream. Ideally, the liquid used to dissolve the sample is the same as the mobile phase. For analytical work, the concentration is normally in the 1 $\mu g/\mu L$ (1 mg/mL) range. For preparative work, the concentration would be greater. If the sample is not sufficiently soluble in the solvent being used for the chromatography, a solvent should be chosen that is *less* polar than the eluting solvent. If a more polar solvent is used, the chromatography may be severely disrupted. A less polar solvent should be used even if a relatively large volume must be used. In this case, the sample will be concentrated in the first part of the column and then chromatographed in the usual way.

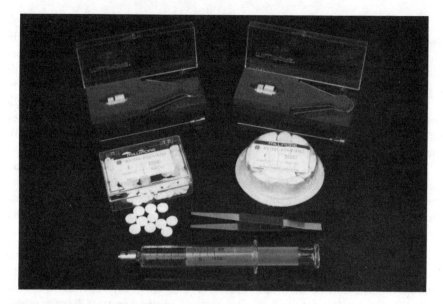

Figure 6.17 Syringe mounted sample filtration system.   (Reproduced through the courtesy of Millipore Corp.)

A known compound that will not be retained by the column may be added to the sample to establish the dead volume of the column and make possible the determination of k' values as discussed above. An example of this is uracil as shown in Figure 6.7. A known material may also be added to a sample in known amounts to serve as an internal standard for quantitative analysis of mixture components.

The sample *must* be filtered before it is placed in the system. Small microfiltration units are available that may be used with a hypodermic syringe for this purpose. Such a syringe is shown in Figure 6.17.

The actual sample size varies widely and depends on the size of column, the detector system available, the type of chromatography being carried out (analytical or preparative) and the ease of the separation. For analytical separations on small columns, samples are in the range of 10^{-5} g or lower per gram of column packing. For preparative work, samples on the order of 10^{-3} g or more per gram of packing are used.[1] For example, a semi-preparative column 10 mm i.d. by 300 mm in length can separate as much as 1 g of sample into broad fractions by overloading the column.

Injection Systems

Sample application to an HPLC column is much like GC in that the sample must be injected into a flowing stream against a back pressure that may be quite high, and it must be introduced in a narrow plug to minimize band broadening. In addition, the method used for the injection must be convenient, reproducible, and be able to cope with pressures up to 6000 psi.

As in GC, the sample may be injected through a rubber or plastic septum with a syringe and needle as long as the back pressure is less than 1500 psi. Actually, even with higher pressures, the pump can be turned off to reduce the pressure and restarted after injection without too much distortion of the chromatography; this is called stop-flow injection. The main problem with a septum is that particles tend to break off of it and clog the system.

Actually, most sample injections are now carried out using an ingenious sliding valve system such as the one shown in Figure 6.18. In the loading position on the right of the figure, the sample solution is loaded into the sample loop with a syringe. The size of the loop (its volume) determines the size of the sample, since the solution is normally injected until it comes out of the overflow. Loops are available in many different sizes (2-5000 μL). Note that while the sample is being loaded into the loop, the flow from the pump to the column is not disturbed. When the valve is rotated to the inject position on the right of the figure, the carrier liquid is diverted to pick up the contents of the sample loop and carry it into the column.

After use, the syringe, the sample loop, its overflow tube, and the valve should be carefully cleaned with fresh solvent or the next sample before being used again.

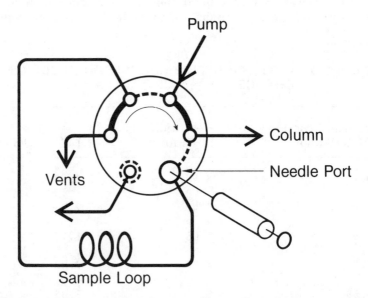

Figure 6.18 Schematic diagram of an injection valve showing the various positions and syringe port. (Reproduced through the courtesy of Rheodyne Corp.)

Other types of valves as well as automated injectors of many types are available.

Columns and Packings

Main Columns. Very little needs to be added to our previous discussion of column packings, both in this chapter and in Chapter 5. We should note that there are several kinds of supports for the bonded phase packings. These, currently all silicas, are: (1) spherical (3-10 μm) totally porous particles, (2) irregularly shaped (5-10 μm) microporous particles, (3) spherical (30-70 μm) macroporous particles, and (4) spherical (30-70 μm) surface porous particles. These latter surface porous spheres consist of a hard core of glass or silica coated with a porous layer of silica and are called pellicular supports. In general, the porous spherical supports are used in most columns and the pellicular particles are used only in guard columns (see below).

As noted previously, bonded phase column packings are prepared by allowing the hydroxy groups on the silica surface to react with various reagents to provide an organic layer on the particles. For steric reasons, it is difficult to produce a packing on that *all* of the hydroxyl groups are reacted, especially with the large reagents such as octyl and octadecyl silyl chlorides (see Table 6.4). The extent to which a surface is completely coated is called its degree of loading. It is possible to treat a partially loaded packing (that is, one that cannot be loaded any more) with a smaller

Table 6.9   RETENTION TIME CHANGES WITH SOLVENT
             CHANGE

Volume Fraction Of Methanol	Retention Time in Minutes	k' (k^o = 1.45 Min)
0.70	3.40	1.34
0.60	5.40	2.72
0.50	9.08	5.16
0.62	4.82	2.32
0.58	5.87	3.05

reagent such as trimethylsilyl chloride to cap or endcap additional hydroxyl groups (many, but not all of them).

Bonded phase packings are difficult to make in a controlled manner on a small scale, and the packing of columns of these materials under pressure is not particularly easy. For these reasons, columns are usually purchased from commercial sources, completely packed and ready to use. Columns can be conditioned for use by passing, through them, several column volumes (7-10 mL) of the solvent that will be used in the chromatogram.

In general, column tubes are made of highly polished stainless steel and are about 3-5 mm inside diameter and 5-30 cm long. Microbore HPLC columns are currently 1 mm x 1 m and packed with 5 μm particles. For low pressure work (less than 300 psi) glass columns can be used *with a good safety shield*.

The efficiency or the number of theoretical plates of any given column can be measured by the method described in the Theoretical Concepts section of Chapter 1. It is only necessary to measure the volume of liquid phase in the column, the retention time of any given solute, and the half-width of the observed peak as it appears on the recorder trace. When peaks are not actually Gaussian (symmetrical), as is often the case, this is called **skew**. When this appears on chromatograms from bonded phase packings, it is generally due to adsorption on free hydroxyl groups that have not reacted with the bonding phase or endcapping reagent.

The column length to be used is directly related to the complexity of the sample to be separated and the efficiency of the column packing. A simple mixture can often be completely separated with a 5 cm column with a high mobile phase flow rate. A more complex mixture might require a 15-25 cm column for complete separation.

The column temperature is generally not as critical in HPLC as in GC. However, if highly reproducible retention volumes are needed, the temperature should be controlled. An increase of temperature, perhaps to 40-50°C, will usually sharpen the peaks.

Guard Columns. Any sample that is to be chromatographed may consist of three parts. One part is material that may be insoluble in the solvent to be used for injection; one part may be soluble or suspendable but will stick in the first part of the column and will not move; and the third part will actually undergo chromatography. Ideally, one would have only the third, since the first two will clog and destroy the column for reuse. The insoluble part can be removed by careful filtration (see above). The soluble or suspendable but non-moving part is usually removed with a guard column.

Guard columns are short, usually 5 cm, columns that are filled with a packing similar (5 or 10 μm) to the main column packing and are placed between the sample injection system and the main column. There are essentially three types. One can be purchased along with the main column.

Another type is a dry repackable column filled with pellicular packing. Because of the nature and price of this packing, the columns are cheap, easily packed, and have little back pressure. Finally, pre-packed cartridges containing 10 μm particles (to produce less back pressure) are available, which fit into a reusable holder.

The guard column should be replaced frequently to maintain the integrity of the main, and much more expensive, column.

Detectors

Little information needs to be given in addition to the previous discussion. Table 6.10 is more complete than Table 6.3 as far as *types* of detectors are concerned, but has less detail on the more common types. Two of these, mass spectrometry and infrared detection, will be discussed in more detail later. In summary, only the refractive index detectors are really universal in any sense, but they are less sensitive.

Clean detector cells are critical to accurate detection. The cells are normally cleaned with concentrated nitric acid followed by washing with water and then the common organic solvents.

Signal Handling

As the solutes leave the column, their concentration is measured in the stream by the detector and the resulting signal (actually voltage) is fed to a recorder. The data on the recorder trace can be viewed from a qualitative or a quantitative point or as a guide on when to cut fractions for a preparative separation. The manipulation of these data is exactly like that used for GC and is discussed thoroughly in Chapter 2.

Fraction Collectors

When the compounds separated by HPLC are to be collected and isolated, it is often convenient to use some type of fraction collecting apparatus, such as the automatic fraction collector shown in Figure 5.8. When a microprocessor-controlled system is in use, the processor can be instructed as to when and how the fractions should be collected.

6.5 SPECIAL TECHNIQUES

Recycle Chromatography

One normally thinks of chromatography as a single pass technique. Thus, the solutes are normally passed once through the system in TLC and classic column separations. However, in HPLC, the columns are reusable and reusable *immediately* so that, with an appropriate valve/tubing ar-

Table 6.10   HPLC DETECTORS AND THEIR PROPERTIES

Detector	Detection Sensitivity	Gradient Sensitivity	Selectivity
Ultraviolet	10^{-10} g/mL	None	Broad capabilities
Refractive index	10^{-7} g/mL	High	Broadest capabilities
Electrochemical	10^{-12} g/mL	High	Oxidizable/ reducible compounds
Conductivity	10^{-8} g/mL	None	Ion detection
Fluorescence	11^{-11} g/mL (8 femtograms)	None	Fluorescent compounds
Flame photometric	10^{-8} g/mL P 10^{-7} g/mL S	High	Element detector
Inductively coupled plasma	10^{-9} g/mL	None	Broad element detection
Mass spectrometer	10^{-9} g/sec	None	Broadly used, even for high MW
Infrared spectrophotometry	10^{-6} g/mL	High[a] or None[b]	Molecular characterization of >100,000 compounds
Light scattering	10^{-4} g/mL	High	Gives M_w of polymers
Viscometer	10^{-6} g/mL	High	Gives M_v of polymers
Post column reaction	10^{-10} g/mL	None	Change lower sensitivity to high sensitivity

[a] *Stop-flow or solvent subtraction methods being used.*
[b] *Solvent removal method being used.*

rangement, the solutes can be recirculated back through the pump to the beginning of the column (or another column). Often this circular flow can be continued until a desired separation has occurred or until one obtains an overlap from one cycle to the next. The problem with this process is that there is always a certain amount of dead volume in the cycle when the solvent stream is flowing through the tubing, valves, and, in particular, the pump. During this time the solute bands can become significantly broadened.

Recycle chromatography can also be used with two identical columns on an alternate basis, two different columns on an alternate basis, and different columns used an unequal number of times. The cycling can be continued until the mixture is spread out to occupy one full column.

Fraction Removal and Heart Cutting

In a recycling system, one can remove any portion of the stream at any given time and continue to recycle what is left. This is sometimes called **heart cutting** and can be used in two ways as shown in Figure 6.19.

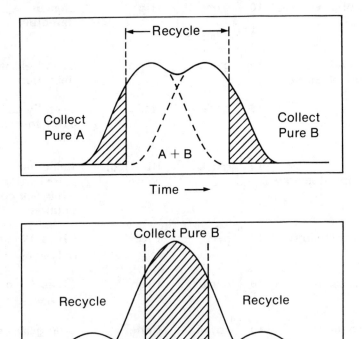

Figure 6.19 Two different modes of heart cutting.

Multi-Dimensional HPLC

This mode of HPLC utilizes **column switching** and most often columns of differing polarity and type. One can, of course, use any number of different columns in series as long as the volume of the connections between them is kept low to prevent back mixing and as long as all of the columns are compatible with the solvent being used. If too many columns are used, however, an intolerable back pressure may build up.

HPLC/Mass Spectrometry

The application of mass spectrometry to the effluent of a chromatographic column permits one to measure the mass spectrum of each of the components or, for that matter, any portion of the stream, be it a pure component or a mixture. This type of data can be used for the detection of solutes and, in particular, for their qualitative identification. This was discussed in Chapter 2 and is an especially important and easy technique for GC, where the mobile phase is an inert gas.

In HPLC, the application of mass spectrometry is complicated by the presence of the liquid mobile phase that must be eliminated before the spectrum can be measured. Commercial instrumentation of various types is now available, but an in-depth discussion of the methods of solvent removal is beyond the scope of this book.

HPLC/Infrared Spectrophotometry

In principle, infrared spectrophotometry can be used for the detection and identification of solutes in much the same manner as mass spectrometry. In this case complications arise from two facts. First, the solvents used have numerous and strong absorptions and are present in overwhelming amounts as is the case in HPLC/mass spectrometry. Second, a normal infrared spectrum requires time for its measurement. Stop-flow techniques have been used in that the effluent from a column is allowed to flow through an IR cell and the flow is stopped while a spectrum is measured. The spectrum of the solvent can be subtracted using a double beam and reference sample system. This, however, allows some back-mixing while the flow is stopped. In order to circumvent this latter problem, a Fourier transform IR (FTIR) can be used. Such instruments are somewhat more expensive than the dispersive type ($30,000-$80,000 or more) but measure spectra very rapidly and can be used directly on a flowing stream. In this case the solvent background can be subtracted electronically with pertinent information stored in the memory of a computer.

Methods have also been developed that permit HPLC/IR to be carried out in an efficient manner by complete solvent removal. One method deposits the samples on KBr pellets and another technique thermosprays the sample on a thin polymer or metal film that can be examined by IR. The FTIR allows the minor background of the KBr or the film to be subtracted or ignored when reflective IR is used. The KBr is limited to non-aqueous systems, but the thermospray method can handle any solvent or solvent mixture.

Quantitative HPLC

Quantitative HPLC is mainly a matter of treating the data generated by a detector in an appropriate manner. This was discussed in detail for GC in Chapter 2 and will not be repeated here. Three cautions might be noted. As in GC, quantitative data are not accurate when columns have been overloaded. Second, the lack of a universal detector might cause problems. Finally, as discussed previously, the detector does not respond equally to all solutes so that the concentration of each solute can be measured *only* using a calibration for that specific solute.

Preparative HPLC

In Chapter 3, we noted that LLC had a low intrinsic capacity for amounts of solute, due mainly to the relatively small amount of stationary liquid phase present on the column. On bonded phase columns with high surface loading, this problem is somewhat alleviated and preparative HPLC is both possible and practical, although many chromatographers are using both flash chromatography and medium pressure systems (see Chapter 5).

Analytical HPLC can be scaled up for preparative work in two ways. First, multiple small injections can be made, much as in preparative GC, as discussed in Chapter 2. The more common way, however, is to use large diameter columns and high solvent throughput, up to 10 or 20 mL/min. Commercial semi-preparative and preparative HPLC columns are available from several manufacturers. The semi-preparative column can be used for samples up to 1 g and the preparative column for 5 g or more. (See an excellent chapter on preparative LC in reference 1.)

One difficulty in analytical HPLC is partially solved with preparative work. The detector need not be as sensitive, for larger amounts of solute are present and quantitative data are not required. A refractive index detector is satisfactory for most purposes. When UV detectors are used, a problem sometimes arises because so much solute is coming from the column that it exceeds the capacity of the detector. Special thin cells in the detector can help the situation, but more often some type of stream splitting is required. Some sort of fraction collecting system is desirable.

Some of the methods discussed above, such as recycling and heart cutting, are most useful in HPLC. For example, one may deliberately overload a column and heart cut as shown in Figure 6.19 to collect a pure fraction. Depending upon the value of the solutes, one could then recycle as shown in the figure to obtain pure fractions of all three solutes. The possibilities for fractionation, heart cutting, and recycling are almost infinite.

6.6 POLYMER CHARACTERIZATION

The use of polymers and plastics in our society is ubiquitous, and the manufacture of these materials involves the major portion of our chemical industry. Furthermore, many biochemical materials, such as proteins, peptides, lignin, polysaccharides, and even some lipids, are polymeric in nature. Since they are high molecular weight materials with low volatility and limited solubilities, they have been difficult to characterize. For example, what is the molecular weight (or degree of polymerization) of a sample? And, how narrow or broad a range of molecular weights is present in a given sample? It is possible to determine the **average molecular weight** by several methods, such as viscosity, vapor pressure and light scattering techniques, but the determination of the range of molecular weights is quite difficult. Data on this matter can readily be obtained from HPLC methods as long as the polymers are soluble in a mobile phase. Actually, HPLC was developed largely as an answer to this problem, although ion exchange chromatography has been used extensively for the separation of proteins and peptides.

Thus far in this book we have discussed chromatography only in terms of adsorption to a solid stationary phase or as solubility in a bonded stationary phase. While these methods can be applied to liquids or polymers, the results are not very satisfactory. Polymer characterization is usually carried out on a stationary phase consisting of a porous material with a closely *controlled pore size.* Discrimination between solute polymers is based upon the apparent *size* of the molecules in solution and, under proper conditions, *nothing else.* This means that chromatographic properties are directly related to size and molecular weight and chromatography can be used to determine these values for numerous polymers. Since this method was originally developed using cross-linked dextran gels (polymers of glucose) and aqueous solvents, the method has been called **gel filtration chromatography** (GFC). A more descriptive term that we will use is **size exclusion chromatography** (SEC).

The phenomena underlying SEC actually differ from other types of chromatography in two ways. First, the mobile phase has little to do with the chromatography, so that different solvents of the same solvating power give similar results. In a sense, the mobile phase in SEC is more like the gas in GC in that it can serve simply as a neutral medium from which solute molecules may enter the stationary phase.

The second way involves the manner in that a controlled pore size packing interacts with solute. At first glance, one might conclude that SEC should not be a very efficient process. For example, if a packing had a certain specific pore size, one might conclude that all of the solute molecules larger than the pore size would be immediately eluted off the column and that *all* of the solutes having a smaller size than the pore size would go into the pores, stay there, and not be eluted at all. The first supposition is quite true and it must be recognized that all of the solutes above a certain size will come off the column in the dead volume as one broad fraction.

The second supposition is not true and represents the crux of SEC. When several solutes can enter a pore of a given size, they will enter at *different rates* and remain there for *different times,* with the smaller solutes moving in faster and staying there longer. Thus, smaller molecules will spend less time in the mobile phase, move more slowly on the column, and come out after larger molecules. This is the reverse situation from most other chromatographic systems where larger molecules elute more slowly than smaller molecules (other things being equal). In summary, a given packing will fractionate solutes having a size small enough to enter its pores.

As stated, the original packings for SEC were cross-linked polymers of glucose, which were prepared by cross-linking naturally occurring dextrans in a controlled fashion (the various Sephadexes), and water was the mobile phase. These gels are still widely used, but a number of other packings have also been developed (see Table 6.11). Some of these are inorganic, such as the silicas and glasses. In this case, the surface hydroxyl groups are usually blocked or capped by various organic bondings, as discussed previously. The various cross-linked polystyrenes are synthetically prepared and cross-linked in a controlled fashion to produce the desired pore size (or type) and gel-particle diameter. In this case, each material is available in **gel permeation chromatography** (GPC) packings. The gel packings come in a series of pore sizes/types and two gel diameters, 5 and 10 μm. These packings are used only for samples soluble in organic solvents.

The apparatus and general techniques of SEC are quite similar to those of HPLC. It is more common, however, to use several columns in series to achieve difficult separations. These multi-column arrangements may have packings of different pore size to give the overall system the ability to separate samples of widely differing molecular weights.

Figure 6.20 shows the separation of a mixture of polyester oligomers (small polymers) derived from the reaction of a diphenol and a diacid chloride. The packing is this case is not a regular SEC material, but a modified silica (amino, see Table 6.4) with a controlled pore size of about 100 Ang. Note that the smaller molecules having molecular weights below about 2500 can be resolved into discrete peaks. For really high molecular weight polymers, such as those over 500,000 or 1,000,000, only a general distribution curve is obtained, such as the one shown in Figure 6.21. The average molecular weight of a polymer is usually determined by comparing its chromatographic properties (retention volume) with those of polymers

Table 6.11   TYPES OF COMMERCIAL EXCLUSION PACKINGS

Type[a]	Trade Name[b]	Particle Shape[c]	Use[d]
Polystyrene	Styragel	SP	O
Polystyrene	PL Gel	SP	O
Polystyrene	Shodex A	SP	O
Polystyrene	TSK Type H	SP	O
Porous glass	Corning	IR	B
Porous glass	BioGlas	IR	B
Silica	LiChrospher	SP	B
Silica	Zorbax	SP	B
Silica	TSK Type SW	SP	A
Silica	Bondagel	SP	B
Polydextran	Sephadex	IR	A
Polydextran	Sephadex LH-20	IR	O
Polyacrylamide	Biogel	IR	A
Agarose	Biogel A	IR	A
Agarose	Sephadex	IR	A

[a] Polymer forming the cross-linked network
[b] See bibliography for more information
[c] SP = Spherical; IR = Irregular
[d] O = Organic; B = Biological; A = Aqueous

(usually polystyrenes) of known molecular weight. Figure 6.22 shows such a calibration curve for a column set. The broadness of a curve, such as that shown in Figure 6.21, is a measure of the molecular weight distribution in a sample, or polydispersivity. Additional information on this important topic can be obtained from books listed in the Bibliography.

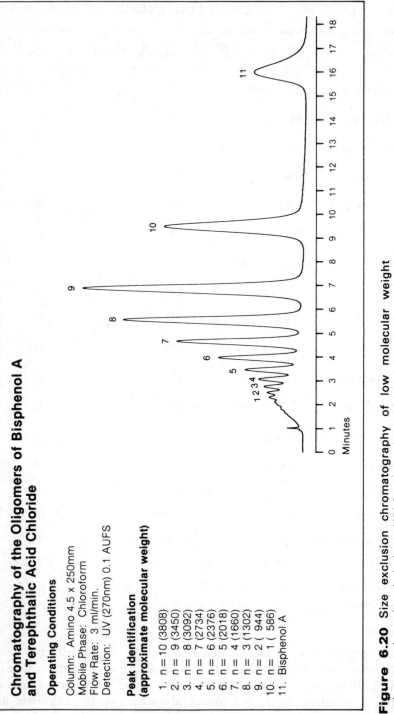

Chromatography of the Oligomers of Bisphenol A and Terephthalic Acid Chloride

Operating Conditions

Column: Amino 4.5 x 250mm
Mobile Phase: Chloroform
Flow Rate: 3 ml/min.
Detection: UV (270nm) 0.1 AUFS

Peak Identification
(approximate molecular weight)

1. n = 10 (3808)
2. n = 9 (3450)
3. n = 8 (3092)
4. n = 7 (2734)
5. n = 6 (2376)
6. n = 5 (2018)
7. n = 4 (1660)
8. n = 3 (1302)
9. n = 2 (944)
10. n = 1 (586)
11. Bisphenol A

Figure 6.20 Size exclusion chromatography of low molecular weight polymers using a bonded phase HPLC column in the SEC mode. Note that the higher molecular weight fractions elute first.

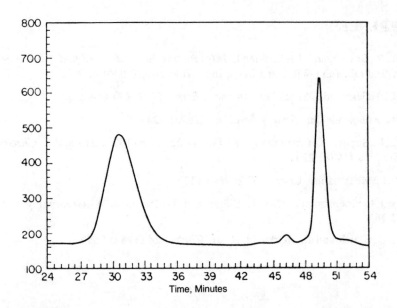

Figure 6.21 Size exclusion chromatography of a styrene/n-butyl methacrylate copolymer showing the broadness of the molecular weight at 30.2 min and the peak broadening in the peaks for the monomers at 45.4 and 48.6 min.

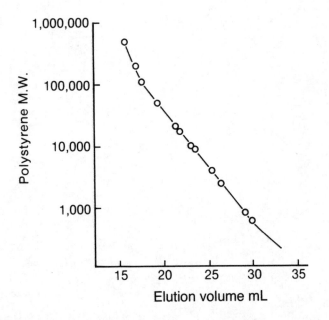

Figure 6.22 Typical calibration curve for a GPC column set.

REFERENCES

[1] L.R. Snyder and J.J. Kirkland, *Introduction to Modern Liquid Chromatography,* 2nd ed., John Wiley and Sons, Inc., New York **(1979)** p. 277.

[2] L. Halasz and I. Sebestian, *Angew. Chem. 81* **(1969)** 464.

[3] L. Rohrschneider, *Anal. Chem. 45* **(1973)** 1241.

[4] L.R. Snyder, *J. Chromatog. 92* **(1974)** 223 and (1974) 223 and *J. Chromatog. Sci. 16* **(1978)** 223.

[5] J.L. Meek, *Anal. Chem. 52* **(1980)** 1370.

[6] P.J. Schoenmakers, H.A.H. Billiet and L. DeGalan, *J. Chromatog. 218* **(1981)** 261.

[7] J.L. Glajch and J.J. Kirkland, *Anal. Chem. 54* **(1982)** 2593.

GLOSSARY

Adsorption Chromatography - Separation utilizing a solid surface as a stationary phase with a liquid or gas moving phase, LSC in this text.

Affinity Chromatography - Liquid Solid Adsorption chromatography in which chemospecific adsorption is utilized and the development involves the chemical displacement of the adsorbed molecules.

Anion Chromatography - see Ion Exchange Chromatography.

Anticircular Thin Layer Chromatography - Thin Layer Chromatography with the origin on an outer ring and development occurring by the mobile phase being delivered to the exterior and flowing inward to produce a series of circular bands.

Ascending Chromatography - Paper or Thin Layer Chromatography in which the mobile phase moves up through the paper or layer by capillary action.

Binary Gradient Chromatography - Liquid Chromatography in which the mobile phase is a two-solvent mixture, the proportions of which are changed during the run.

Bioaffinity Chromatography - see Affinity Chromatography.

Bonded Phase Chromatography - Gas or Liquid Chromatography in which the stationary phase is chemically bonded to the support or to the walls of the capillary column.

Capillary Chromatography - Gas or High Performance Liquid Chromatography in which the column is less than to 0.5 mm in diameter.

Cation Chromatography - see Ion Exchange Chromatography.

Centrifugal Chromatography - Chromatography in which the mobile phase is forced outward in the thin layer by spinning in a centrifuge.

Circular Thin Layer Chromatography - Thin Layer Chromatography carried out with the origin as a spot at the center of the layer and the developing liquid applied to the origin to give a series of circular bands.

Column Chromatography - Liquid Chromatography involving a gravity flow of a liquid mobile phase and, usually, a solid stationary phase.

Countercurrent Chromatography - Liquid Chromatography utilizing liquid-liquid concepts in a process based on countercurrent extraction in which the mobile phase in droplet form rises or falls through an immiscible solvent in a sequence of chambers (columns).

Critical State Chromatography - Separation involving the gas or liquid mobile phase in the critical state; see also Supercritical Fluid Chromatography.

Descending Chromatography - Paper Chromatography in which the mobile phase flows down the paper.

Displacement Chromatography - Separation where the liquid or gas mobile phase displaces the components from the solid stationary phase by being more strongly adsorbed.

Dry Column Chromatography - Chromatography in which the adsorbent is packed dry; the sample is adsorbed on adsorbent and is placed on top of the column; and the mobile phase is added, followed by normal column chromatography. Subsequently, the developed column is allowed to dry and the column is cut in pieces to obtain the fractions.

Electro-Chromatography - Liquid Chromatography combined with electrophoresis, with two electrodes passing current through the mobile phase to assist and/or cause the separation.

Electron Exchange Chromatography - Ion Exchange Chromatography in which the separation depends on the ease of electron transfer between a solute and the solid stationary phase.

Eluent Chromatography - see Elution Chromatography.

Elution Chromatography - Separation in which the components of a mixture are separated by their differences in retention by the solid (LSC) or liquid (LLC) stationary phase, and flow (elute) out of the column.

Emulsion Chromatography - Separation in which the mobile phase is an emulsion of a liquid in a liquid.

Exclusion Chromatography - see Size Exclusion Chromatography.

Extraction Chromatography - Liquid Chromatography in which the mobile phase is an organic liquid and the stationary phase is aqueous and an extraction process occurs.

Flash Chromatography - Liquid Chromatography in which the mobile phase is pushed rapidly through a short, large diameter column packed with an adsorbent of controlled particle size.

Flat Bed Chromatography - see Thin Layer Chromatography.

Foam Chromatography - Separation using either a liquid or gas mobile phase and a solid stationary phase prepared by foaming (while cross-linking) a polymer in a column.

Frontal Chromatography - Separation in which the liquid or gas solute/solvent mixture is the mobile phase and separation occurs by selective adsorption on the solid stationary phase.

Gas Adsorption Chromatography - see Gas Chromatography.

Gas Chromatography - Separation utilizing a gas as the mobile phase and either a solid (LSC) or a liquid on a support (LLC) as the stationary phase.

Gas Liquid Chromatography - see Gas Chromatography.

Gas Partition Chromatography - see Gas Chromatography.

Gas Solid Chromatography - see Gas Chromatography.

Gel Chromatography - see Size Exclusion Chromatography.

Gel Filtration Chromatography - Size Exclusion Chromatography in which an aqueous mobile phase is used with an appropriate hydrophilic gel phase.

Gel Permeation Chromatography - see Size Exclusion Chromatography.

Gradient Liquid Chromatography - Liquid Chromatography in which the composition of the mobile phase is slowly changed, usually making it more polar.

High Efficiency Chromatography - see High Performance Liquid Chromatography.

High Performance Liquid Chromatography - Liquid Chromatography carried out with a stationary phase chemically bonded to a fine, narrow distribution support and a liquid mobile phase that is forced to flow at a controlled rate with high pressure.

High Performance Thin Layer Chromatography - Thin Layer Chromatography on layers of higher quality adsorbent (smaller and narrow particle range) which results in an enhanced separation.

High Pressure Liquid Chromatography - see High Performance Liquid Chromatography.

High Speed Liquid Chromatography - see High Performance Liquid Chromatography.

Hydrodynamic Chromatography - A form of Size Exclusion Chromatography in which the materials (polymers) being separated are suspended in the mobile phase (colloidal), and the stationary phase is a non-porous polymeric particle.

Ion Chromatography - Ion Exchange Chromatography in which a second ion exchange column (the suppressor column) is used to exchange the cations or anions associated with the sample to form the poorly ionized acid of the eluent and permit conductivity detection.

Ion Exchange Chromatography - Separation of ions involving an ionic aqueous mobile phase (acidic, basic, or neutral buffer) and a solid stationary phase having cationic or anionic sites at the surface which can exchange cations or anions from the mobile phase.

Ion Interaction Chromatography - see Ion Pair Chromatography.

Ion Pair Chromatography - Liquid Chromatography in the reversed phase mode using, as the mobile phase, a buffer and an added counter ion of opposite charge to the sample.

Ion Suppression Chromatography - Liquid Chromatography with the pH adjusted to suppress ionization of the sample.

Isocratic Liquid Chromatography - Liquid Chromatography in which the mobile phase is constant in composition, but often involves a mixture of solvents.

Isothermal Chromatography - GC in which the temperature of the column is maintained at one specific temperature.

Linear Ideal Chromatography - The most direct and simplest description of chromatography; does not normally occur.

Linear Non-Ideal Chromatography - Chromatography in which the peaks (bands) broaden because of diffusion effects and non-equilibrium conditions.

Liquid Chromatography - Separation utilizing a liquid mobile phase and either a solid (LSC) or a chemically bonded or adsorbed liquid (LLC) stationary phase.

Liquid Exclusion Chromatography - see Size Exclusion Chromatography.

Liquid Liquid Chromatography - see Liquid Chromatography.

Liquid Solid Chromatography - see Liquid Chromatography.

Mass Chromatography - Gas Chromatography using a gas density balance as a detector which gives molecular weights of the peaks.

Medium Pressure Liquid Chromatography - Liquid Chromatography in which the pressure used is 50-150 psi instead of the 500-6000 psi used for High Performance Liquid Chromatography.

Molecular Sieve Chromatography - Gas Solid Chromatography in which the adsorptive stationary phase is a molecular sieve, and the separation is based on the adsorption of the gas molecules being separated.

Multi-Dimensional Chromatography - Chromatography in which the sample is separated in two or more distinctive gas, liquid, or thin layer chromatographic operations without isolation.

Multiple Chromatography - Thin Layer or Paper Chromatography in which more than one mobile phase is used in series to develop the separation.

Multiplex Chromatography - Gas or liquid chromatography in which samples are continuously injected.   The individual chromatograms are identified with computer techniques involving deconvolution.

Non-Linear Ideal Chromatography - Gas or Liquid Solid Chromatography in which the peaks develop sharp fronts and diffuse rear boundaries (tailing).

Non-Linear Non-Ideal Chromatography - Gas or Liquid Solid Chromatography in which the peaks exhibit diffuse front (fronting) and rear boundaries (tailing).

Normal Phase Chromatography - Liquid Chromatography in which the mobile phase is less polar (hydrocarbon) than the stationary phase (alumina, silica).

Orthogonal Chromatography - Multi-Dimensional Chromatography in which the methods used involve more than one type of separation phenomenon, e.g., partition and size exclusion chromatographies.

Paper Chromatography - Separation utilizing a liquid mobile phase migrating by capillary action and a liquid stationary phase (usually water) suspended in the cellulosic fibers of the paper.

Partition Chromatography - Separation utilizing a liquid stationary phase with either a liquid or a gas mobile phase.

Planar Chromatography - see Thin Layer Chromatography.

Plane Chromatography - see Thin Layer Chromatography.

Plasma Chromatography - A separation technique where ions in a gaseous mobile phase are separated by their mobility down a column that contains no stationary phase (at atmospheric pressure).

Preparative Layer Chromatography - see Thick Layer Chromatography.

Programmed Flow Chromatography - Gas or High Performance Liquid Chromatography in which the mobile phase flow is increased in some linear or stesked rate.

Programmed Temperature Gas Chromatography - Gas Chromatography in which the temperature is increased at a linear or stepped rate.

Pseudophase Liquid Chromatography - Liquid Chromatography in which a micellar or cyclodextrin mobile phase is used, resulting in binding, non-binding, and anti-binding between solute and the mobile phase.

Pulse Chromatography - see Multiplex Chromatography.

Pyrolysis Gas Chromatography - Gas Chromatography in which an insoluble sample (polymer) is rapidly heated (300-900°C) in the injector to form volatile fragments that are separated.

Reaction Chromatography - Gas or High Performance Liquid Chromatography in which a chemical reaction (derivative formation, oxidation, reduction, thermal change, etc.) occurs in the injector or column to give a more volatile, more easily separable, or more easily detectable sample.

Recycle Chromatography - Gas or Liquid Chromatography in which the separation requires sending the sample through a column or columns a number of times using valves.

Reversed Phase Chromatography - Liquid Chromatography in which the mobile phase is more polar (water based) than the bonded or adsorbed stationary phase liquid.

Size Exclusion Chromatography - The separation of compounds (usually polymers) of higher molecular weights on the basis of molecular size and shape using a liquid mobile phase and a highly porous stationary phase (often highly cross-linked polymeric gels) in which the smaller molecules spend a greater amount of time in the pores than the larger molecules to give the achieved resolution.

Steric Exclusion Chromatography - see Size Exclusion Chromatography.

Supercritical Fluid Chromatography - Liquid Chromatography in which the mobile phase is a gas that is kept in the liquid phase by keeping the pressure of the gas above its supercritical point and the stationary phase is either a solid or bonded liquid.

Ternary Gradient Chromatography - Liquid Chromatography in which the concentrations of a three-solvent mobile phase are changed according to some linear or step program during the run.

Thermal Chromatography - Gas Chromatography in which the compounds to be analyzed are freed from a non-volatile solid sample by controlled heating at 25-300°C; see also Pyrolysis Gas Chromatography.

Thick Layer Chromatography - Thin Layer Chromatography on adsorbent layers that are thick enough (0.5-2 mm) to permit the preparative separation of larger amounts of sample.

Thin Layer Chromatography - A separation utilizing a liquid mobile phase migrating by capillary action through a solid (LSC) stationary phase deposited in a uniform thickness on a flat supporting surface.

Two-Dimensional Chromatography - (1) Thin Layer or Paper Chromatography in which development occurs first in one direction and then in a second direction, usually 90°, with a second solvent.   (2) Gas or High Performance Liquid Chromatography in which two different columns are used to permit Gas to Gas, Gas to Liquid, Liquid to Liquid, or Liquid to Gas column switching.

Vacancy Chromatography - A separation using a UV absorbing mobile phase in which the components being separated are not absorbing, thus a peak is a UV vacancy.

Vapor Phase Chromatography - see Gas Chromatography.

SUPPLIERS

The following sections contain most of the suppliers and manufacturers of chromatographic equipment for the United States. A more complete and up-to-date listing can be found in the annual Laboratory Guide, published by the American Chemical Society through the journal *Analytical Chemistry*, in the February issue of the *Journal of Chromatography*, and in the Buyers Guide edition of *American Laboratory*.

GENERAL CHROMATOGRAPHY SUPPLIES

ALLTECH ASSOCIATES, 2051 Waukegan Road, Deerfield, IL 60015.
ANALABS, INC., P.O. Box 501, New Haven, CT 06473.

J.T. BAKER CHEMICAL CO., 222 Red School Lane, Phillipsburg, NJ 08865.

KONTES GLASS CO., P.O. Box 729, Vineland, NJ 08360.

SUPELCO, INC., Supelco Park, Bellefonte, PA 16823.

A.H. THOMAS CO., Vine Street at Third, Philadelphia, PA 19105.

UNIVERSAL SCIENTIFIC, INC., 2070 Peachtree Industrial Court, Suite 101, Atlanta, GA 30341.

GAS CHROMATOGRAPHY - INSTRUMENTS

ANALYTICAL INSTRUMENT DEVELOPMENT INC., Route 41 and Newark Road, Avondale, PA 19311.

ANTEK INSTRUMENTS, INC., 6005 North Freeway, Houston, TX 77076.

CARLE INSTRUMENTS, INC., 1200 Knollwood Circle, Anaheim, CA 92801.

GOW-MAC INSTRUMENT CO., P.O. Box 32, Bound Brook, NJ.

HEWLETT-PACKARD CO., 3000 Hanover Street, Palo Alto, CA 94304.

HNU SYSTEMS, INC., 30 Ossipee Road, Newton, MA 02164.

PACKARD INSTRUMENT CO., INC., 2200 Warrenville Road, Downers Grove, IL 60515.

PERKIN-ELMER CORP., Main Avenue, Norwalk, CT 06856.

SHIMADZU SCIENTIFIC INSTRUMENTS, INC., Oakland Ridge Industrial Center, 9147-H Branch Road, Columbia, MD 21045.

SPECTRA-PHYSICS, 3333 North First Street, San Jose, CA 95134.

TRACOR INSTRUMENTS, 6500 Tracor Lane, Austin, TX 78721.

VARIAN INSTRUMENT GROUP, 611 Hansen Way, Palo Alto, CA 94303.

GAS CHROMATOGRAPHY - COLUMNS

ALLTECH ASSOCIATES, INC., 2051 Waukegan Road, Deerfield, IL 60015.

ANALABS, 80 Republic Drive, North Haven, CT 06473.

CHROMPACK INC., P.O. Box 6795, Bridgewater, NJ 08807.

DEXSIL CHEMICAL CORP., 295 Treadwell Street, Hamden, CT 06514.

HEWLETT-PACKARD CO., 3000 Hanover Street, Palo Alto, CA 94304,

J & W SCIENTIFIC INC., 3871 Security Park Drive, Rancho Cordova, CA 95670.

QUADREX CORPORATION, P.O. Box 3881, Amity Station, New Haven, CT 06525.

SUPELCO, INC., Supelco Park, Bellefonte, PA 16823.

ULTRA SCIENTIFIC, 1 Main Street, Hope, RI 02831.

HIGH PERFORMANCE LIQUID CHROMATOGRAPHY - EQUIPMENT

BECKMAN INSTRUMENTS, INC., 2500 Harbor Blvd., Fullerton, CA 92634.

DUPONT ANALYTICAL INSTRUMENTS, McKean Bldg., Concord Plaza, Wilmington, DE 19898.

FIATRON SYSTEMS INC., 6651 N. Sidney Place, Milwaukee, WI 53209.

HEWLETT-PACKARD CO., 3000 Hanover Street, Palo Alto, CA 94304.

IBM INSTRUMENTS, INC., P.O. Box 332 - Orchard Park, Danbury, CT 06810.

JASCO INCORPORATED, 218 Bay Street, P.O. Box 1499, Easton, MD 21601.

KRATOS ANALYTICAL INSTRUMENTS, 24 Booker Street, Westwood, NJ 07675.

LDC/MILTON ROY CO., P.O. Box 10235, Riviera Beach, FL 33404.

MICROMERITICS INSTRUMENT CORP., 5680 Goshen Springs Road, Norcross, GA 30093.

PERKIN-ELMER CORP., Main Avenue, Norwalk, CT 06856.

RAININ INSTRUMENT CO., Mack Road, Woburn, MA 01801.

RHEODYNE, INC., P.O. Box 996, Cotati, CA 94928.

SPECTRA-PHYSICS, 3333 North First Street, San Jose, CA 95134.

TRACOR INSTRUMENTS, 6500 Tracor Lane, Austin, TX 78721.

VALCO INSTRUMENTS CO., P.O. Box 55603, Houston, TX 77055.

VARIAN ASSOCIATES, 611 Hansen Way, Palo Alto, CA 94303.

WATERS ASSOCIATES, 34 Maple Street, Milford, MA 01757.

HIGH PERFORMANCE LIQUID CHROMATOGRAPHY – COLUMNS

ALLTECH ASSOCIATES, INC., 2051 Waukegan Road, Deerfield, IL 60015.

J.T. BAKER CHEMICAL CO., 222 Red School Lane, Phillipsburg, NJ 08865.

BECKMAN INSTRUMENTS, 2500 Harbor Blvd., Fullerton, CA 92634.

BIO-RAD LABORATORIES, 2200 Wright Avenue, Richmond, CA 94804.

BROWNLEE LABS, INC., 2045 Martin Avenue, Santa Clara, CA 95050.

DUPONT ANALYTICAL INSTRUMENTS, McKean Bldg., Concord Plaza, Wilmington, DE 19898.

EM SCIENCE, 480 Democrat Road, Gibbstown, NJ 08027.

IBM INSTRUMENTS, INC., P.O. Box 332 - Orchard Park, Danbury, CT 06810.

JONES CHROMATOGRAPHY INC., P.O. Box 14425, Columbus, OH 43214.

MCB REAGENTS, 480 Democrat Road, Gibbstown, NJ 08027.

PERKIN-ELMER CORP., Main Avenue, Norwalk, CT 06856.

POLYMER LABORATORIES INC., Box 1581, Stow, OH 44224.

SEPARATIONS GROUP, THE, P.O. Box 867, Hesperia, CA 92345.

SHANDON SOUTHERN INSTR. INC., 515 Broad St., Drawer 43, Sewickley, PA 15143.

SUPELCO, INC., Supelco Park, Bellefonte, PA 16823.

VARIAN ASSOCIATES, 611 Hansen Way, Palo Alto, CA 94303.

WATERS ASSOCIATES, 34 Maple Street, Milford, MA 01757.

WHATMAN INC., 9 Bridewell Place, Clifton, NJ 07014.

THIN LAYER CHROMATOGRAPHY - EQUIPMENT/SUPPLIES

ALLTECH ASSOCIATES, INC., 2051 Waukegan Road, Deerfield, IL 60015.

ANALABS, 80 Republic Drive, North Haven, CT 06473.

ANALTECH, INC., 75 Blue Hen Drive, Newark, DE 19711.

APPLIED ANALYTICAL INDUSTRIES (CAMAG), Route 6, Box 55, New Hanover County Airport, Wilmington, NC 28405.

J.T. BAKER CHEMICAL CO., 222 Red School Lane, Phillipsburg, NJ 08865.

BRINKMANN/SYBRON CORP., Cantaigue Road, Westbury, NY 11590.

EASTMAN KODAK CO., 343 State Street, Rochester, NY 14650.

EM SCIENCE, 480 Democrat Road, Gibbstown, NJ 08027.

GELMAN SCIENCES INC., 600 Wagner Road, Ann Arbor, MI 48106.

KONTES GLASS CO., P.O. Box 729, Vineland, NJ 08360.

REGIS CHEMICAL CO., 8210 Austin Avenue, Morton Grove, IL 60053.

SCHLEICHER & SCHUELL, INC., Keene, NH 03431.

SUPELCO, INC., Supelco Park, Bellefonte, PA 16823.

COLUMN CHROMATOGRAPHY - EQUIPMENT/SUPPLIES

ACE GLASS CO., 1430 N.W. Blvd., Vineland, NJ 08360.

ACE SCIENTIFIC SUPPLY CO., INC., 40-A Cotters Lane, P.O. Box 1018, East Brunswick, NJ 08816.

AMICON, INC., 17 Cherry Hill Drive, Danvers, MA 01923.

BIO-RAD LABORATORIES, 2200 Wright Avenue, Richmond, CA 94804.

BUCHLER INSTRUMENTS, INC., 1327 16th Street, Fort Lee, NJ 07024.

ICN PHARMACEUTICALS, INC. (WOELM), 26201 Miles Road, Cleveland, OH 44128.

BIBLIOGRAPHY

General Texts on Chromatography

Chromatographic Methods, R. Stock and C.B.F. Rice, Chapman and Hall, London, England **(1974)**.

Chromatography and Allied Methods, O. Mikes, Halsted Press, New York **(1979)**.

Chromatographic Systems: Maintenance and Troubleshooting, J.Q. Walker, M.T. Jackson, Jr., and J.B. Maynard, Academic Press, New York **(1977)**.

CRC Handbooks on Chromatography: Drugs **(1981)**, *Carbohydrates* **(1982)**, *Phenols and Organic Acids* **(1982)**, *Polymers* **(1982)**, *General Data and Principles* **(1972)**, CRC Press, Boca Raton, FL.

Theory and Mathematics of Chromatography, A.S. Said, Dr. Alfred Huethig, New York **(1981)**.

Chromatographic Methods in Inorganic Analysis, G. Schwedt, Dr. Alfred Huethig, New York **(1981)**.

Users Guide to Chromatography, Regis Chemical Co., Morton Grove, IL **(1976)**.

Dynamics of Chromatography, Principles and Theory, J.C. Giddings, Marcel Dekker, Inc., New York **(1965)**.

Texts on Gas Chromatography

Modern Practice of Gas Chromatography, R.L. Grob, Wiley-Interscience, New York **(1977)**.

Gas Chromatographic Detectors, D.J. David, Wiley-Interscience, New York **(1974)**.

Gas Chromatography with Glass Capillary Columns, 2nd Ed., W. Jennings, Academic Press, New York **(1982)**.

Comparison of Fused Silica and Other Glass Columns in Gas Chromatography, W. Jennings, Dr. Alfred Huethig, New York **(1981)**.

Basic Gas Chromatography, H.M. McNair and E.J. Bonelli, Varian Aerograph, Walnut Creek, CA **(1968)**.

Ancillary Techniques of Gas Chromatography, L.S. Ettre and W.H. McFadden, Eds., R.E. Krieger, Huntington, NY **(1978)**.

Quantitative Analysis by Gas Chromatography, J. Novak, Marcel Dekker, Inc., New York **(1975)**.

Texts on Thin Layer Chromatography

Thin Layer Chromatography, J.M. Bobbitt, Reinhold, New York **(1963)**, out of print.

Practice of Thin Layer Chromatography, J.C. Touchstone and M.F. Dobbins, 2nd Ed., Wiley-Interscience, New York **(1983)**.

High Performance Thin Layer Chromatography, A. Zlatkis and R.E. Kaiser, Eds., Elsevier, New York **(1977)**.

Thin Layer Chromatography, K. Randerath, Academic Press, New York **(1963)**.

Quantitative Thin Layer Chromatography, J.C. Touchstone, Ed., Wiley-Interscience, New York **(1973)**.

Thin Layer Chromatography, J.G. Kirchner, Wiley-Interscience, New York **(1967)**.

Texts on High Performance Liquid Chromatography

Instrumental Liquid Chromatography, N.A. Parris, 2nd ed., Elsevier, New York **(1984)**.

Microcolumn High Performance Liquid Chromatography, P. Kucera, Elsevier **(1984)**.

Techniques in Liquid Chromatography, C.F. Simpson, Wiley, New York **(1983)**.

Introduction to High Performance Liquid Chromatography, 2nd ed., R.J. Hamilton and P.A. Sewell, Methuen **(1982)**.

Maintaining and Troubleshooting HPLC Systems, D.J. Runsen, Wiley, New York, **(1981)**.

Introduction to Modern Liquid Chromatography, 2nd ed., L.R. Snyder and J.J. Kirkland, Wiley-Interscience, New York **(1979)**.

Instrumentation for High Performance Liquid Chromatography, J.F.K. Huber, Elsevier, New York **(1978)**.

The LDC Basic Book on Liquid Chromatography, S.B. Schram, Milton Roy Co., St. Petersburg, FL **(1980)**.

Liquid Chromatography Detectors, R.P.W. Scott, Elsevier, New York **(1977)**.

Liquid Chromatography Detectors, T.M. Vickrey, ed., Marcel Dekker Inc., New York **(1983)**.

Reversed Phase HPLC: Theory, Practice and Biomedical Applications, A.M. Krstulovic and P.R. Brown, Wiley-Interscience, New York **(1982)**.

Optimization in HPLC, R.E. Kaiser and E. Oelrich, Dr. Alfred Huethig, New York **(1981)**.

Modern Size Exclusion Liquid Chromatography, W.W. Yau, J.J. Kirkland and D.D. Bly, Wiley-Interscience, New York **(1979)**.

Practical Liquid Chromatography, R.W. Yost, L.S. Ettre and R.D. Conlon, Perkin-Elmer Corporation, Norwalk, CT **(1980)**.

INDEX

(Titles in which Chromatography is capitalized indicate a listing in the GLOSSARY: the companies are listed in SUPPLIERS.)